Table of Contents

Sections: **Day:**

2-Step Problems 1-5

3-Step Problems 6-15

Problems with Exponents 16-20

Problems with Parentheses 21-35

Nested Parentheses 36-50

5-step Problems or more 51-60

Negative Numbers 61-70

Fraction Bars 71-80

(Answer Key in Back)

ISBN: 978-1-63578-322-3

Current contact information can be found at
www.HumbleMath.com
www.LibroStudioLLC.com

Cover image credit: Helen Stebakov/Shutterstock.com

Day 1

2-Step Problems

Name: ____________________

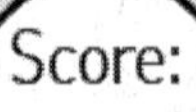

① $13 + 7 \cdot 3$

② $9 \cdot 2 - 13$

③ $32 \div 8 + 4$

④ $29 - 24 \div 2$

⑤ $20 \div 4 \cdot 9$

⑥ $15 - 7 \cdot 2$

⑦ $14 + 8 \div 2$

⑧ $45 \div 9 \cdot 2$

⑨ $30 - 20 \div 5$

⑩ $24 + 6 \cdot 6$

Day 2

2-Step Problems

Name: ____________________

Score:

① 45 - 40 ÷ 8

② 28 + 8 ÷ 4

③ 3 + 2 · 9

④ 6 · 4 - 7

⑤ 10 ÷ 2 · 5

⑥ 10 - 2 · 4

⑦ 16 + 4 ÷ 2

⑧ 24 ÷ 3 · 2

⑨ 12 - 4 ÷ 4

⑩ 2 + 2 · 3

Day 3

2-Step Problems

Name: ____________________

① $5 + 5 \cdot 6$

② $3 \cdot 5 - 4$

③ $49 \div 7 + 10$

④ $19 - 21 \div 3$

⑤ $15 \div 15 \cdot 5$

⑥ $17 - 18 \div 3$

⑦ $33 + 1 \cdot 1$

⑧ $9 \div 3 \cdot 6$

⑨ $50 - 35 \cdot 1$

⑩ $13 + 12 \div 4$

Day 4

2-Step Problems

Name: ____________________

Score:

① $28 - 16 \div 8$

② $16 \cdot 4 + 20$

③ $4 + 1 - 2$

④ $18 \div 9 \cdot 2$

⑤ $28 - 3 \cdot 9$

⑥ $2 + 7 \cdot 18$

⑦ $9 \cdot 3 + 5$

⑧ $16 \div 8 - 2$

⑨ $20 - 15 \div 3$

⑩ $40 + 2 - 28$

Day 5

2-Step Problems

Name: ____________________

Score:

① $30 \cdot 4 + 5$

② $5 - 3 \div 3$

③ $8 + 1 \cdot 10$

④ $16 \div 8 + 27$

⑤ $50 \cdot 3 \div 15$

⑥ $44 - 40 \div 4$

⑦ $17 + 20 - 7$

⑧ $16 \cdot 8 - 50$

⑨ $5 - 2 \cdot 1$

⑩ $6 + 20 - 20$

Day 6

3-Step Problems

Name: ____________________

Score:

① $6 + 10 - 18 \div 3$

② $13 + 7 \cdot 15 \div 5$

③ $28 - 12 \cdot 2 + 6$

④ $32 \div 8 \cdot 5 + 4$

⑤ $26 + 8 \cdot 24 \div 6$

⑥ $20 \div 4 \cdot 11 + 3$

⑦ $4 + 8 \div 2 \cdot 3$

⑧ $14 + 14 \div 7 - 2$

⑨ $3 \cdot 12 + 4 \cdot 10$

⑩ $10 \cdot 2 - 16 \div 4$

Day 7

3-Step Problems

Name: ____________________

① $3 \cdot 5 - 15 \div 5$

② $48 - 40 + 2 \cdot 4$

③ $17 - 6 \div 3 \cdot 1$

④ $5 + 37 \cdot 24 - 23$

⑤ $22 - 3 \cdot 5 + 20$

⑥ $3 + 9 \div 3 \cdot 3$

⑦ $9 \cdot 6 - 38 + 5$

⑧ $6 - 3 + 8 \cdot 2$

⑨ $7 \cdot 5 - 3 \div 3$

⑩ $1 + 49 \cdot 4 \div 4$

Day 8

3-Step Problems

Name: ____________________

① $28 + 6 \cdot 3 - 1$

② $9 \cdot 6 \div 2 + 4$

③ $36 \div 6 + 15 \cdot 2$

④ $11 - 5 \cdot 2 + 3$

⑤ $22 + 18 - 40 \div 8$

⑥ $35 - 5 \div 1 \cdot 7$

⑦ $50 \div 50 \cdot 20 + 20$

⑧ $7 \cdot 16 + 21 - 11$

⑨ $9 + 2 - 1 \cdot 8$

⑩ $13 - 5 \cdot 0 + 34$

Day 9

3-Step Problems

Name: ____________________

① $7 \cdot 13 - 14 \div 7$

② $27 + 0 \cdot 6 - 6$

③ $41 - 20 \div 10 \cdot 9$

④ $8 \div 8 + 3 \cdot 3$

⑤ $2 \cdot 22 + 17 - 16$

⑥ $5 + 39 - 36 \div 9$

⑦ $44 - 2 \div 2 + 4$

⑧ $12 \div 6 \cdot 37 - 33$

⑨ $5 + 3 - 15 \div 5$

⑩ $6 \cdot 10 + 23 - 5$

Day 10

3-Step Problems

Name: ____________________

Score:

① $36 \div 9 + 11 \cdot 3$

② $18 - 2 \cdot 6 + 48$

③ $12 \cdot 1 - 6 + 2$

④ $4 + 14 \div 1 \cdot 8$

⑤ $9 - 3 \cdot 3 + 6$

⑥ $36 \div 12 + 24 - 11$

⑦ $17 \cdot 3 - 19 + 25$

⑧ $16 + 16 \div 16 - 8$

⑨ $41 - 12 + 1 \cdot 0$

⑩ $2 \cdot 10 - 18 \div 2$

Day 11

3-Step Problems

Name: ____________________

① $1 + 17 \cdot 12 - 8$

② $35 \div 5 - 2 \cdot 3$

③ $6 \cdot 1 + 6 \div 2$

④ $26 - 12 \div 6 + 4$

⑤ $43 + 8 - 16 \cdot 3$

⑥ $9 \div 9 \cdot 2 - 2$

⑦ $1 \cdot 23 + 6 - 1$

⑧ $14 - 3 \cdot 2 + 17$

⑨ $50 \div 25 - 1 \cdot 1$

⑩ $13 - 15 \div 3 + 5$

Day 12

3-Step Problems

Name: ______________

Score:

① $7 \cdot 7 - 34 \div 2$

② $24 - 4 \div 2 \cdot 1$

③ $3 + 9 \cdot 9 - 3$

④ $45 \div 5 + 0 - 0$

⑤ $20 - 4 \div 2 \cdot 8$

⑥ $11 \cdot 6 - 26 + 6$

⑦ $8 + 19 \cdot 21 \div 7$

⑧ $14 \div 14 + 3 - 2$

⑨ $8 - 4 \div 1 + 1$

⑩ $26 \cdot 2 - 37 + 8$

Day 13

3-Step Problems

Name: ________________

① $50 - 26 + 8 \div 4$

② $28 + 18 \div 9 - 28$

③ $36 \div 6 - 2 + 1$

④ $1 \cdot 9 + 13 - 15$

⑤ $13 - 8 \cdot 1 + 2$

⑥ $41 + 6 - 8 \cdot 3$

⑦ $1 \cdot 25 \div 5 + 1$

⑧ $33 \div 3 + 27 \cdot 3$

⑨ $41 + 16 - 16 \div 4$

⑩ $19 - 7 \cdot 2 + 7$

Day 14

3-Step Problems

Name: ____________________

Score:

① $6 \cdot 9 - 50 + 2$

② $19 - 18 \cdot 3 \div 3$

③ $42 + 49 \div 7 - 40$

④ $38 \div 2 - 1 \cdot 5$

⑤ $8 \cdot 7 + 16 \div 8$

⑥ $15 \quad 14 \cdot 1 + 4$

⑦ $10 - 4 \div 1 \cdot 1$

⑧ $17 + 6 - 30 \div 5$

⑨ $11 \cdot 2 + 25 - 22$

⑩ $44 \div 4 \cdot 13 + 7$

Day 15

3-Step Problems

Name: ____________________

① $35 - 14 \div 2 \cdot 5$

② $12 + 37 \cdot 24 \div 8$

③ $4 \cdot 26 - 50 + 40$

④ $9 \div 9 + 5 - 2$

⑤ $0 + 1 \cdot 1 \div 1$

⑥ $6 \cdot 9 \div 9 + 25$

⑦ $6 \cdot 6 - 13 + 8$

⑧ $27 \div 3 + 24 \cdot 4$

⑨ $48 - 28 + 5 \cdot 43$

⑩ $15 + 5 \cdot 7 - 7$

Day 16
Exponents

Name: ____________________

Score:

① $6 + 7^2 - 18 \div 3$

② $24 + 3^2 \cdot 4 \div 2$

③ $21 - 6 \cdot 3 + 8^2$

④ $6^2 \div 3 \cdot 2 + 2^2$

⑤ $31 + 4^2 \div 2 \cdot 6$

⑥ $20 \div 4 \cdot 3^2 - 5^2$

⑦ $230 - 8 \div 2^2 \cdot 7$

⑧ $9^2 + 24 \div 8 - 2$

⑨ $3 \cdot 5 + 4^2 \cdot 10$

⑩ $5^2 \cdot 2 - 40 \div 5$

Day 17

Exponents

Name: ____________________

① $40 \div 10 + 1 \cdot 4^2$

② $5^2 \cdot 4 \div 20 + 50$

③ $31 + 6 - 9^2 \div 3$

④ $49 - 2^2 \cdot 8 + 2$

⑤ $1 \cdot 2^2 + 3^2 - 2$

⑥ $5^2 + 16 \div 16 - 10$

⑦ $18 \div 9 - 1^2 + 1$

⑧ $2^2 \cdot 2^2 - 6 + 5$

⑨ $7 + 6 \cdot 6 \div 3^2$

⑩ $32 - 12 + 6^2 \cdot 1$

Day 18

Exponents

Name: ____________________

Score:

① $8^2 - 9 \cdot 7 + 8^2$

② $2^2 + 3^2 - 18 \div 6$

③ $2 \cdot 5^2 + 41 - 34$

④ $16 \div 4^2 + 1 \cdot 7^2$

⑤ $14 + 39 - 6^2 \div 3$

⑥ $48 - 18 \div 3^2 + 1$

⑦ $2 \cdot 2 + 20 - 4^2$

⑧ $27 \div 3^2 - 1 \cdot 2$

⑨ $25 - 10 \div 2 + 1^2$

⑩ $1^2 + 5 \cdot 2^2 - 1$

Day 19

Exponents

Name: ____________________

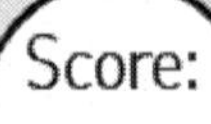

① $26 + 7^2 - 37 \cdot 1^2$

② $85 - 15 \cdot 2^2 + 2$

③ $49 \div 7^2 + 6^2 - 22$

④ $5^2 \cdot 3 \div 15 - 5$

⑤ $20 - 4 \cdot 5 + 4^2$

⑥ $11 + 2^2 - 35 \div 7$

⑦ $6^2 \cdot 1 \div 2 - 4^2$

⑧ $54 - 27 + 3^2 \cdot 3$

⑨ $14 + 1^2 - 18 \div 3^2$

⑩ $36 \div 12 \cdot 5^2 - 50$

Day 20

Exponents

Name: ____________________

Score:

① $50 - 7^2 \cdot 1 + 4$

② $2^2 \cdot 7 + 5^2 - 20$

③ $27 + 2 - 7^2 \div 7$

④ $10^2 \div 10 \cdot 1 + 5^2$

⑤ $6 \cdot 3^2 + 13 - 12$

⑥ $32 - 8 \div 1^2 \cdot 4$

⑦ $2^2 + 38 - 2 \cdot 4^2$

⑧ $7 \cdot 8 + 6^2 \div 6^2$

⑨ $19 - 16 \div 1^2 + 2$

⑩ $18 \div 3^2 \cdot 25 - 24$

Day 21

Parentheses

Name: ____________________

Score:

① $(6 + 10) \div 8 - 1$

② $4 + 4 \cdot (15 - 5)$

③ $25 - (6 + 2^2 \cdot 3)$

④ $(4^2 - 9) \cdot (8 + 2)$

⑤ $(22 + 8)2 \div 6$

⑥ $20 \div (2 \cdot 2) + 3$

⑦ $(4 + 20) \div (3^2 - 3)$

⑧ $14 \div (9 + 7 - 2)$

⑨ $(3 + 5 \cdot 4)10$

⑩ $32 \div 8(6 + 4)$

Day 22

Parentheses

Name: ____________________

Score:

① $6 + 6^2 \div (9 - 3)$

② $2^2 + 4(30 - 5^2)$

③ $100 - (3 \cdot 5 + 7^2)$

④ $(4^2 - 14)^2 (10 - 3^2)$

⑤ $(2 + 8)3^2 \div 3$

⑥ $5^2 \div (7 - 2) + 14$

⑦ $150 \div (8 - 3)^2 \cdot 2$

⑧ $18 \div (6^2 - 5 \cdot 6)$

⑨ $(8^2 + 3 \cdot 4)10$

⑩ $(8 + 4^2) \div (5 + 3)$

Day 23

Parentheses

Name: ____________________

Score:

① $(2 \cdot 4^2)\ (27 - 26)$

② $9\ (3 + 7^2) - 5$

③ $2^2 + 6\ (25 \div 5^2)$

④ $(3^2 + 7 \cdot 3^2)\ 2$

⑤ $(47 - 46) + 6^2 \cdot 6$

⑥ $(10^2 + 10) \div (10 + 1^2)$

⑦ $49 + (45 \div 3^2 - 2^2)$

⑧ $(14 \div 7) - 1 \cdot 1^2$

⑨ $(18^2 \div 18) - (3^2 + 8)$

⑩ $35 + (9^2 - 7^2) + 6$

Day 24

Parentheses

Name: ____________________

Score:

① $8^2 - (9^2 \div 3) + 5^2$

② $(11 + 1^2) \div 3 + 6$

③ $32 + (4 \cdot 3^2 \div 2)$

④ $(42 + 7) - (1^2 \cdot 7^2)$

⑤ $(4^2 - 14 + 4^2) : 6$

⑥ $1 \cdot 5^2 \div (50 - 5^2)$

⑦ $(34 - 6) \div 2^2 - 2$

⑧ $(10^2 \div 50) + (3 \cdot 10)$

⑨ $36 \div (5^2 - 7 \cdot 3)$

⑩ $(7 \cdot 4 - 11 + 12)^2$

Day 25

Parentheses

Name: ____________________

Score:

① $35 \div 7\,(50 - 7^2)$

② $10(9^2 \div 9^2 - 1)$

③ $(50 + 3^2 \cdot 2) \div 2^2$

④ $8^2 \div (28 - 26) + 1^2$

⑤ $(18 - 4^2)\,(5 + 12)$

⑥ $(50 - 6^2 + 1) - 3$

⑦ $10(22 \div 11) + 5^2$

⑧ $18 - (3^2 \cdot 9 \div 9)$

⑨ $(6^2 \div 18 - 2) + 27$

⑩ $(50 \cdot 4) - 11 + 9^2$

Day 26
Parentheses

Name: ____________________ Score:

① $7^2 \cdot 1 - (22 + 9)$

② $16 \div 8(3 \cdot 2^2)$

③ $28 - (4 \cdot 4 \div 4)^2$

④ $(6 + 5)^2 - (1 + 6)^2$

⑤ $(9^2 \cdot 1^2 \div 3^2) + 10$

⑥ $(38 - 28) \cdot 6^2 \div 36$

⑦ $296 - (21 + 10 \cdot 5^2)$

⑧ $12 + (6^2 \div 2^2 - 9)$

⑨ $7^2 - (8 + 7) \cdot 3$

⑩ $(2 \cdot 7)(18 - 17)$

Day 27

Parentheses

Name: ____________________

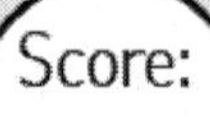

① $(16 - 10)^2 + (13 + 20)$

② $2^2 \div (1 + 1 \cdot 1)$

③ $5 + (18 \div 3)^2 \cdot 3$

④ $(16 \cdot 5^2 - 8) + 70$

⑤ $(11 + 12)^2 - (12 \div 12)^2$

⑥ $37 - (20 \cdot 4) \div 8$

⑦ $10^2 \div (5 + 3 \cdot 5)$

⑧ $27 + 7(2 \cdot 1)$

⑨ $30 + (39 - 9) \div 10$

⑩ $(15 \cdot 6 - 40) + 7$

Day 28

Parentheses

Name: ____________________

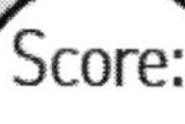

① $29 + 36 - (10^2 \div 50)$

② $(3^2 - 6) - (1 + 1^2)$

③ $(4 \cdot 2^2 + 3) - 4^2$

④ $(81 \div 9 + 6^2) \cdot 1$

⑤ $(7^2 - 49) + (2 \cdot 1)^2$

⑥ $(2 + 8)(8 + 9)$

⑦ $10^2 \cdot (5 - 10 \div 5)$

⑧ $(4 + 3 \cdot 8) \div 7$

⑨ $6^2 + (6^2 - 4 \div 4)$

⑩ $(2^2)(81 \div 3^2) - 20$

Day 29

Parentheses

Name: ____________________

① $3^2 - (1 \cdot 2^2 + 1^2)$

② $5 \cdot 9 + (12 - 2^2)$

③ $(18 + 8 - 19)^2 - 12$

④ $(5 - 1)^2 \div (9 - 1)$

⑤ $11 + (5^2 \div 5^2) - 11$

⑥ $49 \div 7 \cdot (6^2 - 5^2)$

⑦ $5(2 + 2) \cdot 3^2$

⑧ $16 + (7^2 \cdot 1^2 - 9)$

⑨ $(9^2 + 3^2 - 50) \div 8$

⑩ $(9 - 2)^2 - (9 \cdot 2)$

Day 30

Parentheses

Name: ____________________

Score:

① $20 \cdot 6^2 - (8 + 1)^2$

② $160 \div (1 + 3 \cdot 5)$

③ $9 + (3 \cdot 3 \div 3)^2$

④ $(35 + 4^2) - (1 + 2^2)^2$

⑤ $(49 - 7^2 + 3) \div 3$

⑥ $(36 \div 6) \cdot 4 - 2^2$

⑦ $19 + (8^2 - 6 \cdot 7)$

⑧ $5^2 + (6 \div 3) - 5^2$

⑨ $(5^2 - 24 + 7) \div 4$

⑩ $(1 \cdot 6)^2 - (4 - 2)^2$

Day 31

Parentheses

Name: ________________

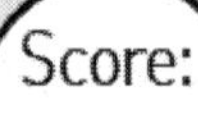

Score:

① $(349 - 7^2)(1 + 2)$

② $(3 + 2)^2 \div (30 - 5^2)$

③ $(6^2 - 1^2 + 5^2) \div 30$

④ $(9 \div 3^2 + 3^2) \cdot 4$

⑤ $(10 - 8)^2 + (2 - 1)^2$

⑥ $(6^2 \div 3) - (4 \cdot 1^2)$

⑦ $45 + (6 \cdot 2^2 + 2^2)$

⑧ $50 \div (36 - 26)(7)$

⑨ $7 + (43 - 14 \div 2)$

⑩ $100(81 \div 9^2 - 1^2)$

Day 32

Parentheses

Name: ____________________

Score:

① $(6 + 4)^2 - (5 \div 1)^2$

② $28 - (49 \div 7^2 \cdot 28)$

③ $3 \div (2 - 1)^2 \cdot 3^2$

④ $(4 \cdot 2 - 2)^2 + 2$

⑤ $(5 \cdot 6 - 2) \div (7 - 5)$

⑥ $6 + 12 \div (10 - 7)$

⑦ $(4)\ (7^2 + 1^2 \cdot 3)$

⑧ $27 - (77 \div 11)\ (2)$

⑨ $80 \div (1 + 3) + 10$

⑩ $(7 \cdot 2 - 2) \div 6$

Day 33

Parentheses

Name: ________________

① $18 \div 9 + (4 - 3)^2$

② $3(1 + 7 \cdot 2)$

③ $6^2 + (4^2 - 2 \cdot 3)$

④ $(8 + 0)^2 - (24 + 27)$

⑤ $(35 + 25 - 12) \div 8$

⑥ $150 \div 15 \cdot (4^2 - 2^2)$

⑦ $19 - (3 \div 3 \cdot 1)^2$

⑧ $5 + (6 \div 6) - 2^2$

⑨ $(3 + 15 \div 3)^2 + 16$

⑩ $(7^2 \cdot 4^2) + (30 - 21)$

Day 34

Parentheses

Name: ______________

① $(8 - 8) \cdot (3^2 + 5^2)$

② $28 \div (10 - 3 \cdot 2)$

③ $6 + (9 - 4) \cdot 7$

④ $(28 - 21 + 1) \cdot 3^2$

⑤ $(8^2 + 1^2) - (12 \div 3)$

⑥ $12 \quad (6^2 : 3) : 2$

⑦ $30 + (5 - 3 \div 1)^2$

⑧ $2 + 5(2 \cdot 1)$

⑨ $6^2 + (5^2 - 7) \div 6$

⑩ $(30 - 10 \cdot 2) + 7$

Day 35

Parentheses

Name: ____________________

① $12 \div (2 \cdot 3 - 4)^2$

② $30 - 4^2 \div (14 - 6)$

③ $(9^2 + 23 - 49) \div 5$

④ $(7 + 2)(1 + 4)$

⑤ $8^2 + (81 \div 9) - 7^2$

⑥ $60 \div 3 \cdot (28 \div 7)$

⑦ $8 \div (6 \div 3)^2 \cdot 5$

⑧ $16 + (3 + 2^2 \cdot 9)$

⑨ $(9^2 + 3^2 - 50) \div 2$

⑩ $(6^2 - 24) - (3^2 \cdot 1)$

Day 36

Nested Parentheses

Name: ______________________

Score:

① $((6 + 7^2) \div 5)2$

② $7((40 - 6^2) \div 2)$

③ $4(72 - (4 \cdot 7 + 6^2))$

④ $(8^2 - 59)((3 + 3^2) \div 2)$

⑤ $(4 + (6 - 4)) \div 3$

⑥ $2^2((7 - 2)2)$

⑦ $36 \div ((3 + 3)^2 \div 4)$

⑧ $56 \div (2^2(5 - 3))$

⑨ $(61 - (7^2 + 6))10$

⑩ $(12 - 4) \div ((6 + 2) \div 4)$

Day 37

Nested Parentheses

Name: ____________________

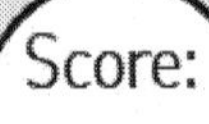

① $((31 + 23)) - (9 \div 3)^2$

② $80 - ((6 + 2)^2 + 2)$

③ $34 - (5^2 + 4^2 \div (1 + 1))$

④ $(3^2 - 2) + (7 \div (4 - 3))^2$

⑤ $((4 \cdot 2 \div 2)^2 - 4) + 4^2$

⑥ $2((34 - 25) + 9^2 - 8^2)$

⑦ $45 - (3 + (6^2 \div 9)^2)$

⑧ $14 + (6^2 + (8 \div 2 + 33))$

⑨ $150 + ((8 - 2)^2 - 3^2)$

⑩ $(4 + (24 - 6)) \div 2$

Day 38

Nested Parentheses

Name: ____________________

Score:

① $42 - ((8 - 5)^2 + 3)$

② $120 \div ((20 - 2) + 12) - 4$

③ $(16 + (5 + 3^2))2$

④ $36 - ((6 - 3)^2 + (36 - 9))$

⑤ $((7^2 + 10 - 2) + 6) \div 9$

⑥ $3((9 + 1) \cdot 2^2 - 5^2)$

⑦ $4^2 \div ((2 \cdot 2) + 2^2)$

⑧ $5((6^2 + 8 \div 2) - 33)$

⑨ $42 \div ((7 - 2)^2 - 19)$

⑩ $(5 + (5 - 1)^2) \div (4 + 3)$

Day 39

Nested Parentheses

Name: ____________________

Score:

① $2(42 + 56 \div (9 - 2))$

② $3((7 + 3) \cdot 5 - 7^2)$

③ $30 \div ((8 - 5)^2 - 4)$

④ $(100 - (5 + 3)^2) \div 9$

⑤ $10((9 - 3 \div 3) - 1)$

⑥ $9((5 - (1 + 1)^2 - 1))$

⑦ $12 + (3(2 + 1))^2$

⑧ $((2 + 4)^2 + 8) \div (7 - 3)$

⑨ $6 \div (100 \div (7^2 + 1))$

⑩ $3(3^2 + 4(3 + 1))$

Day 40

Nested Parentheses

Name: ____________________

Score:

① $(32 - 5^2)((20 + 28) \div 8)$

② $(1 + 7(13 - 8)) \div 3$

③ $5^2 + (4 \div (3 - 2))^2$

④ $((12 - 4 \div 2) - 2)3$

⑤ $((8 - 5)2)^2 - 32$

⑥ $28 \div ((11 - 9) + (2^2 + 1))$

⑦ $(7^2 - 4) \div ((8 + 7) \div 3)$

⑧ $6(9^2 - (8 \cdot 7 + 4^2))$

⑨ $9 + (8^2 - 2(10 + 15))$

⑩ $(50 - (5 + 6 \cdot 2))10$

Day 41

Nested Parentheses

Name: ____________________

① $30 \div (2 + (8 - 4)^2 \div 2)$

② $3^2 + (3^2 - (8 - 5))$

③ $(26 + 16) \cdot 1^2 \div (1 \cdot 1)^2$

④ $((6 \div 2) - 3) + 1^2 - 1^2$

⑤ $(8 - (3 - 2))^2 + 37$

⑥ $120 \div ((20 - 2) + 12) - 1$

⑦ $(5 + 10 - (2 + 6))9$

⑧ $((6 + 6) \div 2)\,(2 + 1)$

⑨ $(4^2 + (38 - 18)) + 4^2$

⑩ $(50 - (45 - 40)) \div (7 - 2)$

Name: ____________________

① $1 - (3^2 \div (14 + 20 - 25))$

② $5((7^2 + 5 \cdot 3) - 33)$

③ $(55 - 10) + (1^2 \cdot (23 - 13))$

④ $(8 - (1 + 3))^2 \div 8$

⑤ $(6 - 1) + ((1 + 2) + 26)$

⑥ $4^2 \div ((12 - 5 \cdot 2) + 6) \cdot 15$

⑦ $(80 + (4^2 \div 8)) - 6$

⑧ $2(42 + 56 \div (9 - 2))$

⑨ $(6 + (6 - 5))^2 + 8$

⑩ $2((70 - 55) + 5^2 - 10)$

Day 43

Nested Parentheses

Name: ____________________

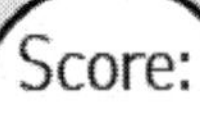

① $(5 + (2 \cdot 14) + (5 \cdot 3^2))$

② $6^2 \div ((7 - 3)3 - 3)$

③ $(5 \cdot (48 \div 4))2 - 15$

④ $(56 \div 7)^2 - ((6 \div 6) + 5^2)$

⑤ $100 - 50 + (150 - (50 + 25))$

⑥ $(22 + (8 \cdot 2^2)) - (4^2 + 16)$

⑦ $(100 - (4^2 + 9 \cdot 8))2$

⑧ $(3 \cdot 2)^2 - (45 \div (7 - 2))$

⑨ $(4 + (30 - 12)) \div 2$

⑩ $2((20 - 14) + 7 - 3^2)$

Day 44

Nested Parentheses

Name: ____________________

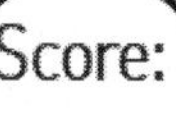

① $56 \div ((6 - 2)^2 \div 2)$

② $((3 \cdot 2)^2 - 16) \div 5$

③ $23 + ((27 \div 9) \cdot 5^2) - 8^2$

④ $(150 \div 50) \cdot (4 + (27 - 23))$

⑤ $((3^2 - 6) \cdot (4 + 2^2)) - 24$

⑥ $6^2 \div (16 \div (3 + 1))$

⑦ $(61 + (8 - 6))10$

⑧ $9 \cdot 3 + (80 \div (16 - 14))$

⑨ $10^2 + ((32 + 68) \div 4)$

⑩ $(72 - (4^2 + 4 \cdot 7)) \div 2$

Day 45

Nested Parentheses

Name: ____________________

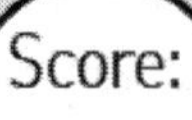

① $44 - 28 + ((8 \div 2)^2 - 13)$

② $50 - ((12 - 7)^2 + 6)$

③ $(10^2 \div (37 - 17)) + 5 \cdot 8$

④ $90 + ((3 \div 1 + 6)^2 + 2)$

⑤ $9 \cdot 6 + (8 - (3 + 4))$

⑥ $48 \div ((18 - 4^2) + (25 - 15))$

⑦ $((20 + 10 - 2) + 8) \div 6$

⑧ $(6 + (12 + 7))5$

⑨ $(7 - 4)\ ((8 + 16) \div 2)$

⑩ $10(2 + (17 - 9))$

Day 46

Nested Parentheses

Name: ____________________

Score:

① $2 + (6 + (3 \cdot 2)^2) \div 7$

② $(12 - (1 + 7))^2 \div 2$

③ $100 - ((45 + 40) - (4 \cdot 2)^2)$

④ $(14 - 6) - (1 + (27 - 22))$

⑤ $(7^2 + (24 - 13)) + 3^2$

⑥ $((15 + 20) \div 5)\ (3 + 4)$

⑦ $(3 + 2)^2 - (10 - (9 - 3))^2$

⑧ $(8 - (3 - 2))^2 + 37$

⑨ $8^2 - 1^2 + (24 \div (7 - 1))$

⑩ $90 \div ((38 + 41) - 7^2)$

Day 47

Nested Parentheses

Name: ____________________

① $(40 - (30 - 8)) \div (7 - 4)$

② $48 \div (1 \cdot (4^2 - 10))$

③ $(5^2 + (3 + 3))^2 - (2 \cdot 2)^2$

④ $4 + (43 - (2 \cdot 3)^2 \div 6^2)$

⑤ $((22 - 17) \cdot 3) - (18 \div 9)$

⑥ $(1 + 2)\,(5 \cdot (5 - 4))$

⑦ $4^2 + (5^2 - (20 - 5))$

⑧ $50 - (10 + 2(10 \div 10))$

⑨ $20 + ((3^2 - 2)^2 - (4 \cdot 2))$

⑩ $(40 - 2(4 + 3^2)) + 34$

Name: ____________________

Score:

① $100 - ((3 \cdot 4 \div 2)^2 + 4^2)$

② $4((5 + 3) \div 2) - 9$

③ $(81 \div (4^2 - (2^2 \cdot 3) + 5))$

④ $(20 - 12)\ ((3 + 7) \div 2)$

⑤ $22 + ((6 \div 2)^2 \quad (1 + 5))$

⑥ $(52 - (13 - 6)) \div (8 + 1^2)$

⑦ $3 + (5^2 - (5 + 40 \div 4))$

⑧ $(8 - (3 - 2))^2 + 37$

⑨ $24 \div ((6 - 4 + 2) \cdot 2)$

⑩ $((1 + 3) \cdot 2)^2 - 34$

Day 49

Nested Parentheses

Name: ____________________

① $((20 + 70) - (3 + 9 \cdot 8)) \div 3$

② $(5 + 10 - (2 + 6))9$

③ $(41 - (14 - 5)) \div (12 - 8)$

④ $(9^2 - (5 + 9 \cdot 7)) + 4^2$

⑤ $(2 \cdot (6 - 1))^2 \div 4$

⑥ $(18 + 22) \div (4 \cdot (34 - 24))$

⑦ $4^2 \div ((5 \cdot 7 + 1) - 20)$

⑧ $((10 - 5) \cdot ((9 \div 3)^2 + 1))$

⑨ $2((2 + 2 - 2)^2 \div 2)$

⑩ $((50 - 5 \cdot 8) + 30)2$

Day 50

Nested Parentheses

Name: ____________________

Score:

① $(4 \cdot 5 - (2 + 14))(5 + 2)$

② $7((4^2 + 10 \div 2) - 10)$

③ $120 \div ((6 + 3 \cdot 3)2)$

④ $37 - (6 \div 3 \cdot (3 + 1^2)^2)$

⑤ $((7 - 5)^2 + (5 \cdot 7)) \div 3$

⑥ $(2(42 - 24)) + 8 : 4$

⑦ $81 \cdot 2 - ((6 \cdot 2)^2 + 6)$

⑧ $43 - ((30 - 9) \div 3 + 2)$

⑨ $(9 + 1)^2 - (15 - (7 - 4))$

⑩ $27 + (12 + 30 \div (3 - 2))$

Day 51

5-Step Problems

Name: ____________________

① $(100 + 50) \div 2 - 3(5 - 1)$

② $(3 + 5) \cdot 2(40 - 4) \div 3^2$

③ $30 - 6 + 4^2 - 3 \cdot 7 + 10$

④ $(6^2 - 9 \cdot 3)\ (8 + 2 \div 2)$

⑤ $(22 + 8)5 \div 25 - 2 \cdot 2$

⑥ $24 \div (40 - 4^2 \cdot 2)\ (3 + 7)$

⑦ $(37 + 5) \div (3^2 - 4 \cdot 4 \div 8)$

⑧ $14 \div 2 + 3 \cdot 8 - 20 \div 4$

Day 52

5-Step Problems

Name: ____________________

① $5 - (1 \cdot 2) + 3 \div (2 - 1)$

② $4 \div 1 \cdot 3^2 \div (2 + 7) - 3$

③ $40 + 27 - (5 \cdot 3) + (8 \cdot 1^2)$

④ $(4 \cdot 2 + 2)^2 \div 20 - 1 + 1$

⑤ $9 - 9 + 1 \cdot 2\,(4 \cdot 3)^2$

⑥ $(8 + 7 \cdot 4 - 2) - 3^2 \div 9$

⑦ $1 \cdot 6 - 2 + (80 \div 40 + 3)$

⑧ $7 \div (6 - 5) + 28 \cdot (5 \div 5)^2$

Day 53

5-Step Problems

Name: ____________________

① $9^2 \cdot 2 \div 81 - 1 + 6^2 - 30$

② $5 + 7 - 8 \cdot 3 \div 24 + 2^2$

③ $13 - (42 - 6 \cdot 7) + 5 \cdot 5$

④ $18 \div 9 + 5 \cdot (4 + 4 - 5)^2$

⑤ $37 \div (10 + 11 \cdot 1 + 16 \cdot 1)$

⑥ $(2 \cdot 2)^2 + 5 - 13 \div 13 + 17$

⑦ $29 + (7 + 8 - 1 \cdot 1)^2 \div 98$

⑧ $(4 - 3)^2 + (9 \div 9)^2 \cdot (1 \cdot 3)$

Day 54

5-Step Problems

Name: ____________________

Score:

① $(9 + 2)^2 - 6 \div 1 \cdot (35 - 18)$

② $8 \cdot 5 + (4 - 2)^2 - 2 \div 1$

③ $49 \div 7^2 + 3 - 18 \div (4^2 + 2)$

④ $28 - 5 \cdot 2 \div 10 + 4 \cdot 5$

⑤ $(23 - 24 \div 8 + 10) \cdot 2 - 5$

⑥ $34 + 15 - (2 \cdot 7 - 1)^2 \div 169$

⑦ $4 \cdot 7 + 4^2 - 195 \div (1^2 + 4)$

⑧ $30 \div 2 - 3 \cdot 2 + 20 \cdot 1$

Day 55

5-Step Problems

Name: ____________________

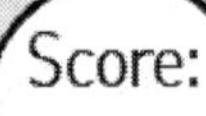

① $100 \div 10 + (8 \cdot 4) - 36 \div 6$

② $(150 + 40 \div 8 - 6 + 8) - 157$

③ $3 \cdot (2 - 1)^2 + (34 \div 17 - 1^2)$

④ $(50 - 9 \cdot 5) - 64 \div 4^2 + 8^2$

⑤ $8 \div 4 \cdot 14 - 16 + 1 \cdot 11$

⑥ $7^2 - (26 + 74) \div 20 \cdot (7 \div 1)$

⑦ $19 + (8 - 8 \div 8 \cdot 2 - 4)^2$

⑧ $5 \cdot (36 \div 9 + 9 - 12) \cdot 5$

Day 56

5-Step Problems

Name: ____________________

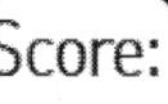

Score:

① $(50 - 30) \cdot (2 + 1)^2 - 81 \div 9$

② $3^2 \div 9 + (5 \cdot 3 - 15) \cdot 3$

③ $26 + 2^2 \cdot 2 - 7^2 \div 49 - 0$

④ $5 \cdot (17 - 25 \div 5) \cdot 1 + 5$

⑤ $10 + 10 - (100 \div 10) + (1 \cdot 3)^2$

⑥ $81 \div 3^2 + 25 \div 5^2 \cdot 18 \div 3$

⑦ $(6^2 \div 3) \cdot (2^2 + 2) - (4 + 3^2)$

⑧ $(15 - 7) \div 8 \cdot 9 - 9 + 1^2$

Day 57

5-Step Problems

Name: ____________________

Score:

① $4 \cdot 4^2 - 64 + 9^2 \div 3^2 - 4$

② $22 + 23 \cdot 10^2 \div 100 - 10 + 17$

③ $(5 - 5)^2 + (3 \cdot 3)^2 \div (3 \cdot 3)$

④ $(36 \div 6 + 8 \cdot 1) \cdot (6 \div 6)$

⑤ $(7 + 5)^2 - (8 \div 2 + 2 \cdot 8)$

⑥ $9 \div (49 - 40) + (3 + 7 \cdot 2)^2$

⑦ $29 - 17 + 3 \cdot 32 \div 96 - 8$

⑧ $(4 \cdot 2 - 2)^2 - (6 - 3)^2 + 36$

Day 58
5-Step Problems

Name: ____________________

Score:

① $(5^2 - 5) \cdot 2^2 \div 5 + (7^2 \div 7)$

② $(3^2 \cdot 4) + (5^2 - 3^2 + 23) - 0$

③ $25 \div 5 + 60 - 5 \cdot 2 + 19$

④ $36 + (7 - 3)^2 \cdot (10 \div 2) \cdot 1$

⑤ $48 - 6 \cdot (1 \div 1 + 7 \cdot 1)$

⑥ $3^2 \cdot 81 \div 3^2 - 26 + 12 - 1^2$

⑦ $6 + (18 - 36 \div 6) + 150 - 50$

⑧ $30 \div 15 + 7 + 5 \cdot 3^2 \div 3$

Day 59

5-Step Problems

Name: ____________________

Score:

① $2^2 + (26 - 10) \cdot 2^2 \div (49 - 41)$

② $45 - 5 \cdot 3^2 + 3^2 - 5^2 \div 5$

③ $4 \cdot 8 - (18 \div 6) + 5 - 5$

④ $(1 \div 1 \cdot 7)^2 + 13 - 3 + 6$

⑤ $6^2 + 11 - (6 - 6^2 \div 36) \cdot 3^2$

⑥ $15 - (6 \cdot 2 + 3) \div (5 - 2)$

⑦ $(3 \div 3) + (4 + 3 \cdot 2)^2 + 2$

⑧ $(5 - 1) \cdot 2 + 2 \div (6 - 4)$

Day 60

5-Step Problems

Name: ________________

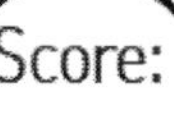

① $(35 - 17) \cdot 1 + (50 + 9 \div 3)$

② $(6 + 4 - 2)^2 \div (4^2 + 4^2 - 30)$

③ $49 \div 7 + 11 - 100 \div 10 \cdot 1$

④ $8 \cdot 64 \div 4 - 2 \cdot 8 - 24$

⑤ $3^2 \cdot (2 - 1 + 2) \cdot (2 \cdot 5)$

⑥ $(20 + 10) \div 5 \cdot 2 \div (2 - 1)^2$

⑦ $32 - 9 \div 3 + 21 \div 7 - 7$

⑧ $2 \cdot 5 - 2^2 + 10 - 4 \cdot 2^2$

Day 61

Negative Numbers

Name: ___________________

Score:

① 10 + (5 - (-8))

② ((8 ÷ (-4)) + (-24))

③ ((-3) + (-9 + 3))

④ (-7 · (-36 ÷ -6))

⑤ ((12 - (-3)) + 6)

⑥ ((-15 + -7) - 8)

⑦ (((-7) + 2) · 9)

⑧ (20 ÷ -5 + (-35))

⑨ (2 · (-24 ÷ 8))

⑩ ((-6) + (-10) + 8)

Day 62

Negative Numbers

Name: ____________________

Score:

① 3 - ((-5) + 9)

② ((-3) – 5) + 6

③ ((-7) + (-4)) - 2

④ (-3) + ((-8) + 12)

⑤ (8 - (-7)) + 5

⑥ ((-6) + 4) + (-1)

⑦ ((-3) + 2) + 9

⑧ (5 - (-4)) + 7

⑨ ((-10) – 4) + (-8)

⑩ 3 + ((-5) + 9)

Day 63

Negative Numbers

Name: ____________________

Score:

① -20 - (-15 + 10)

② (35 - 11) - 23

③ (7) + (-4 + 2)

④ 18 - (41 - (-31))

⑤ ((-16) + (-5)) - 6

⑥ ((-25) - 13 - (-12))

⑦ (17 + (-7)) - 3

⑧ ((-28) + (-18) - 8)

⑨ (3 + (-3) - (-2))

⑩ (24 - (-15) - 13)

Day 64

Negative Numbers

Name: ________________

Score:

① (-36 + (-16) - (-6))

② (5) + (-8 + (-3))

③ 12 - ((-9) - (-7))

④ 47 + ((-13) + 4)

⑤ (1 + (-1) - (-2))

⑥ (6 - 7 + (-9))

⑦ -29 + (14 - (-8))

⑧ (9 - (-3)) - 22

⑨ ((-20) + 10 + (-40))

⑩ ((-3) + (-9 + 3))

Day 65

Negative Numbers

Name: ________________

Score:

① $((8 - 11) + (-6))$

② $4 - ((-9) - 7)$

③ $((-6) + (-10) + 8)$

④ $(5 \cdot ((-81) \div 9))$

⑤ $((8 \div (-4)) + (-24))$

⑥ $(-11 + 21) \cdot (-3)$

⑦ $5 \cdot ((-49) \div (-7))$

⑧ $((-12) \div 6) + 39$

⑨ $((10 \div (-2)) \cdot 10)$

⑩ $(12 - 4 \cdot (-4))$

Day 66

Negative Numbers

Name: ______________

① $((9 \cdot (-3)) \div (-27))$

② $(4 \div (-2) - 36)$

③ $18 + ((-2) \cdot (-5))$

④ $((7 - (-9) + 6))$

⑤ $(-6 \cdot 8 \div 12)$

⑥ $11 - (21 + (-3))$

⑦ $(9 \div 3 - (-31))$

⑧ $((-16) \div 8) \cdot 2$

⑨ $((-26) + 14 - (-14))$

⑩ $(3 +(-6) \div (-3))$

Day 67

Negative Numbers

Name: ____________________

① $((24 - 44) \cdot (-4))$

② $(2 \cdot (-24 \div 8))$

③ $(20 - 40 + (-25))$

④ $((-81) \div (3 - 2))$

⑤ $((-7) + (-7) - 7)$

⑥ $(8 \div (-2)) \cdot 2$

⑦ $(5 \cdot 5 + (-5))$

⑧ $(3 \cdot (-36 \div 6))$

⑨ $((-42 \div 6) + 15)$

⑩ $(13 - ((-4) - 37))$

Day 68

Negative Numbers

Name: ________________

Score:

① $((-18 \div 9) - 7)$

② $(5 \cdot 2 \div -10)$

③ $(4 + (-3) \cdot -6)$

④ $(20 \div -5 + (-35))$

⑤ $((8 - 2) \div -6)$

⑥ $(-7 \cdot (-36 \div -6))$

⑦ $(100 \div -50 - 10)$

⑧ $((17 + (-19)) \cdot 2)$

⑨ $((3 + 9) - (-39))$

⑩ $((-34) + (-13) + 7)$

Day 69

Negative Numbers

Name: ____________________

Score:

① (-1 · 5 · 4)

② ((-5) - ((-2) + (-10)))

③ ((24 ÷ 6) + (-24))

④ ((-0) - ((-50) ÷ -5))

⑤ ((-15 + -7) - 8)

⑥ (37 + 48 + (-17))

⑦ ((1) ((-14) + (-42)))

⑧ (10 ÷ ((-6) - 4))

⑨ ((-11) - 6) ÷ -17

⑩ (7 - (-3) + (-2))

Day 70

Negative Numbers

Name: ________________

Score:

① (((-19) - (-17)) + (-15))

② ((8 · -4) ÷ -16)

③ (3 - 5 · (-6))

④ (((-7) + 2) · 9)

⑤ ((-1) + ((-6) - 7))

⑥ (40 ÷ 4 · -3)

⑦ ((12 - (-3)) + 6)

⑧ ((9 · -2) + (-8))

⑨ ((-0) + ((-29) ÷ 1))

⑩ ((21 ÷ (7)) - (-21))

Day 71

Fraction Bars

Name: ________________

Score:

① $\dfrac{(35 + 45 \div 3)}{5(8 - 6)}$

② $\dfrac{9 - 32 \div 8 - 2}{5^2 - 3 \cdot 7}$

③ $\dfrac{7^2 - 6 \cdot 8}{6 + 35 \div 7}$

④ $\dfrac{2(40 - 4)}{3^2}$

⑤ $\dfrac{2(17 - 6)}{60 \div 5 + 5 \cdot 3}$

⑥ $\dfrac{100 - 5^2 \cdot 3}{4 + 12 \cdot 2}$

⑦ $\dfrac{2^2 \cdot 2}{2(6^2 - 4(5 + 1))}$

⑧ $\dfrac{11 + 3 \cdot 9 - 2}{(53 - 8) \div 5}$

Day 72

Fraction Bars

Name: ____________________

Score:

① $\dfrac{3^2 + 13}{50 - 5^2}$

② $\dfrac{(22 \div 11 \cdot 7) + 4}{44 - (6 \cdot 7)}$

③ $\dfrac{8^2 \div (7 + 9)}{5 \cdot 3 - 2^2}$

④ $\dfrac{11 - 6 + 6}{28}$

⑤ $\dfrac{(13 - 3) + 3^2}{4 \cdot 7 + 8 - 15}$

⑥ $\dfrac{(8^2 + (45 \div 15))}{(39 - 29) \cdot 7}$

⑦ $\dfrac{(2 \cdot 5) - (8 \div 4)}{49 \div 7^2 + 7}$

⑧ $\dfrac{24 - 4^2 \div 4 + 30}{15 + (9 \cdot 4)}$

Day 73

Fraction Bars

Name: ______________ Score:

① $\dfrac{(20 \cdot 2) - 35}{9 + 7 \div 7}$

② $\dfrac{6 + (8 - 4) \cdot 6}{10 \div 5 + 8}$

③ $\dfrac{36 - 10 + 8^2}{15 \cdot 3^2 \div 9}$

④ $\dfrac{4^2 \div 2 \cdot 5 - 7}{(28 - 25 + 7)^2}$

⑤ $\dfrac{9 \cdot (4 - 2)^2}{(30 \div 15 - 2) + 36}$

⑥ $\dfrac{(16 - 6)^2 \cdot 2}{5 \cdot (24 - 14)}$

⑦ $\dfrac{(7 + (7 \cdot 3) \div 21)}{((91 - 10) \div 3^2)}$

⑧ $\dfrac{(30 \div 6 + 2)^2 - 5^2}{3 \cdot 9 - 1}$

Day 74

Fraction Bars

Name: ____________________

Score:

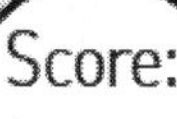

① $\dfrac{((12 \div 6) \cdot 3)}{90 \div (1 \cdot 10)}$

② $\dfrac{25 - 13 + (16 \div 8)}{(15 - 9) \cdot 2^2}$

③ $\dfrac{(4 \cdot 1)^2 \div 2}{5^2 - (8 + 9)}$

④ $\dfrac{18 - 7^2 \div 7 + 3}{8 \cdot 9 \div 2}$

⑤ $\dfrac{(8 \div 2)^2 - 6}{(8 + 7)2 - 25}$

⑥ $\dfrac{(3^2 + 16 - 3^2)}{39 - 10 + 4}$

⑦ $\dfrac{19 \cdot 3 - 49 \div 7}{9 \cdot 9 \div 81}$

⑧ $\dfrac{12 - 10 + 1^2 \cdot 4}{(11 - 7)^2 \cdot 2}$

Day 75

Fraction Bars

Name: ____________________

Score:

① $\dfrac{18 + (5 \cdot 2)^2}{59}$

② $\dfrac{36 \div 12 - 1}{(3 + 2)13}$

③ $\dfrac{(7^2 + 11)4}{(12 - 10)^2}$

④ $\dfrac{(4 + 5)^2}{(9 - 2) \cdot (8 + 5)}$

⑤ $\dfrac{6 \cdot 3 \div 9}{28 \div (4 - 2)^2 + 2}$

⑥ $\dfrac{(18 - (8 + 6))}{49 \div 7}$

⑦ $\dfrac{8 + ((5 \cdot 5) - 8)}{11 - (7 + 3)}$

⑧ $\dfrac{(9^2 - (4 + 2 \cdot 2)^2)}{5((7 - 3)^2 + 2^2)}$

Day 76

Fraction Bars

Name: ____________________

Score:

① $\dfrac{(8 - 6 \div 3)^2}{(2 + 2)^2 \div (2 \cdot 2)}$

② $\dfrac{((4^2 \cdot 2) + 7)}{(6^2 + 3)}$

③ $\dfrac{4 \cdot 5 - 6}{17 + 6 - 9 \div 3}$

④ $\dfrac{12 \div 3 + 26}{(2 \cdot 5)^2 \div 20}$

⑤ $\dfrac{150 - (7 + 4)^2 \cdot 1}{6^2}$

⑥ $\dfrac{6\,(4 - 3)}{3 \div 3(3)}$

⑦ $\dfrac{8^2}{8^2 \cdot 6 \div 96}$

⑧ $\dfrac{22 \div 11 \cdot 6 - 10}{0 + 11}$

Day 77

Fraction Bars

Name: ___________________

Score:

① $\dfrac{6^2 - 7 \cdot 5}{14 \div 7 \cdot 4^2}$

② $\dfrac{8\,(5) \div 10}{5^2}$

③ $\dfrac{((23 + 19) - 34)}{2 \cdot (16 \div 8)^2}$

④ $\dfrac{(24 \cdot 3) \div 3^2}{96 \div (2^2 \cdot 6)}$

⑤ $\dfrac{(21 \div (2 + 5))}{41 + 80 \div 10}$

⑥ $\dfrac{(5)1^2}{((25 - 13) + 7)}$

⑦ $\dfrac{8 \cdot 2^2 - 7^2 \div 49 + 1}{5^2 - 17}$

⑧ $\dfrac{(28 - 16) \cdot 8 - 15}{(6 \div 6 + 8)^2}$

Day 78

Fraction Bars

Name: ______________

Score:

① $\dfrac{21 + (9 \div 3)}{(16 - 13)4}$

② $\dfrac{(6 \div 3 + 2 - 2)^2}{(5 - 4 \cdot 1)^2}$

③ $\dfrac{38 + 23 - 7 \cdot 8}{(8)\ (5)}$

④ $\dfrac{20^2 \div 100}{42 + 8^2 - 24}$

⑤ $\dfrac{2^2 \cdot (4^2 + (8 \div 2))}{((9 \div 3) + 3^2) - 2^2}$

⑥ $\dfrac{39 + 7^2 - 49}{9^2 \div 3 + 27}$

⑦ $\dfrac{8^2 \cdot 5}{32}$

⑧ $\dfrac{(12 - 6) + (12 - 6)}{15 \cdot 2 + 5}$

Day 79

Fraction Bars

Name: ________________ Score:

① $\dfrac{108 \div 6^2 - 1}{3^2 - 5 + 16}$

② $\dfrac{3 \cdot 5 + 2}{((2 + 2)^2 - 8) \cdot 3}$

③ $\dfrac{(4^2 + 11) \div 3^2}{8 \cdot 6^2 - 9^2}$

④ $\dfrac{38 - 33 + 4}{72 \div 8}$

⑤ $\dfrac{((3 \cdot 8) - 15)}{(24 - (10 \div 5) + 11)}$

⑥ $\dfrac{20 \div 4 + 20}{(15 - 10)^2 \cdot 5}$

⑦ $\dfrac{(23 - 13) \cdot 6 \div 2}{5 \cdot 2}$

⑧ $\dfrac{1(13 - 3 + 5)}{(18 \div 2)3}$

Day 80

Fraction Bars

Name: ____________________

Score:

① $\dfrac{9 - 2 \cdot 4}{14 \div 7 \cdot 4^2}$

② $\dfrac{(3 + 3) \div 1 - 2}{5^2 \div 5 + 3}$

③ $\dfrac{10 + (9 - 5)}{21 \cdot 3}$

④ $\dfrac{19 + 13 - 11}{3^2 \cdot (6 + 4)}$

⑤ $\dfrac{4^2 + 7^2}{74 + 8 \div 2 - 5}$

⑥ $\dfrac{(5 + 1) \cdot (11 - 9)}{10 - 3 + 8}$

⑦ $\dfrac{8 \cdot 2^2 - 14}{3^2}$

⑧ $\dfrac{3 + (4 \cdot 4) \div 2}{28 \div 7 + 10}$

Day 1

① $13 + 7 \cdot 3$
$13 + 21$
34

② $9 \times 2 - 13$
$18 - 13$
5

③ $32 \div 8 + 4$
$4 + 4$
8

④ $29 - 24 \div 2$
$29 - 12$
17

⑤ $20 \div 4 \cdot 9$
$5 \cdot 9$
45

⑥ $15 - 7 \cdot 2$
$15 - 14$
1

⑦ $14 + 8 \div 2$
$14 + 4$
18

⑧ $45 \div 9 \cdot 2$
$5 \cdot 2$
10

⑨ $30 - 20 \div 5$
$30 - 4$
26

⑩ $24 + 6 \cdot 6$
$24 + 36$
60

Day 2

① $45 - 40 \div 8$
$45 - 5$
40

② $28 + 8 \div 4$
$28 + 2$
30

③ $3 + 2 \cdot 9$
$3 + 18$
21

④ $6 \cdot 4 - 7$
$24 - 7$
17

⑤ $10 \div 2 \cdot 5$
$5 \cdot 5$
25

⑥ $10 - 2 \cdot 4$
$10 - 8$
2

⑦ $16 + 4 \div 2$
$16 + 2$
18

⑧ $24 \div 3 \cdot 2$
$8 \cdot 2$
16

⑨ $12 - 4 \div 4$
$12 - 1$
11

⑩ $2 + 2 \cdot 3$
$2 + 6$
8

Day 3

① $5 + 5 \cdot 6$
$5 + 30$
35

② $3 \cdot 5 - 4$
$15 - 4$
11

③ $49 \div 7 + 10$
$7 + 10$
17

④ $19 - 21 \div 3$
$19 - 7$
12

⑤ $15 \div 15 \cdot 5$
$1 \cdot 5$
5

⑥ $17 - 18 \div 3$
$17 - 6$
11

⑦ $33 + 1 \cdot 1$
$33 + 1$
34

⑧ $9 \div 3 \cdot 6$
$3 \cdot 6$
18

⑨ $50 - 35 \cdot 1$
$50 - 35$
15

⑩ $13 + 12 \div 4$
$13 + 3$
16

Day 4

① $28 - 16 \div 8$
$28 - 2$
26

② $16 \cdot 4 + 20$
$64 + 20$
84

③ $4 + 1 - 2$
$5 - 2$
3

④ $18 \div 9 \cdot 2$
$2 \cdot 2$
4

⑤ $28 - 3 \cdot 9$
$28 - 27$
1

⑥ $2 + 7 \cdot 18$
$2 + 126$
128

⑦ $9 \cdot 3 + 5$
$27 + 5$
32

⑧ $16 \div 8 - 2$
$2 - 2$
0

⑨ $20 - 15 \div 3$
$20 - 5$
15

⑩ $40 + 2 - 28$
$42 - 28$
14

Day 5

① $30 \cdot 4 + 5$
$120 + 5$
125

② $5 - 3 \div 3$
$5 - 1$
4

③ $8 + 1 \cdot 10$
$8 + 10$
18

④ $16 \div 8 + 27$
$2 + 27$
29

⑤ $50 \cdot 3 \div 15$
$150 \div 15$
10

⑥ $44 - 40 \div 4$
$44 - 10$
34

⑦ $17 + 20 - 7$
$37 - 7$
30

⑧ $16 \cdot 8 - 50$
$128 - 50$
78

⑨ $5 - 2 \cdot 1$
$5 - 2$
3

⑩ $6 + 20 - 20$
$26 - 20$
6

Day 6

① $6 + 10 - 18 \div 3$
$6 + 10 - 6$
$16 - 6$
10

② $13 + 7 \cdot 15 \div 5$
$13 + 105 \div 5$
$13 + 21$
34

③ $28 - 12 \cdot 2 + 6$
$28 - 24 + 6$
$4 + 6$
10

④ $32 \div 8 \cdot 5 + 4$
$4 \cdot 5 + 4$
$20 + 4$
24

⑤ $26 + 8 \cdot 24 \div 6$
$26 + 192 \div 6$
$26 + 32$
58

⑥ $20 \div 4 \cdot 11 + 3$
$5 \cdot 11 + 3$
$55 + 3$
58

⑦ $4 + 8 \div 2 \cdot 3$
$4 + 4 \cdot 3$
$4 + 12$
16

⑧ $14 + 14 \div 7 - 2$
$14 + 2 - 2$
$16 - 2$
14

⑨ $3 \cdot 12 + 4 \cdot 10$
$36 + 4 \cdot 10$
$36 + 40$
76

⑩ $10 \cdot 2 - 16 \div 4$
$20 - 16 \div 4$
$20 - 4$
16

Day 7

① $3 \cdot 5 - 15 \div 5$
$15 - 15 \div 5$
$15 - 3$
12

② $48 - 40 + 2 \cdot 4$
$48 - 40 + 8$
$8 + 8$
16

③ $17 - 6 \div 3 \cdot 1$
$17 - 2 \cdot 1$
$17 - 2$
15

④ $5 + 37 \cdot 24 - 23$
$5 + 888 - 23$
$893 - 23$
870

⑤ $22 - 3 \cdot 5 + 20$
$22 - 15 + 20$
$7 + 20$
27

⑥ $3 + 9 \div 3 \cdot 3$
$3 + 3 \cdot 3$
$3 + 9$
12

⑦ $9 \cdot 6 - 38 + 5$
$54 - 38 + 5$
$16 + 5$
21

⑧ $6 - 3 + 8 \cdot 2$
$6 - 3 + 16$
$3 + 16$
19

⑨ $7 \cdot 5 - 3 \div 3$
$35 - 3 \div 3$
$35 - 1$
34

⑩ $1 + 49 \cdot 4 \div 4$
$1 + 196 \div 4$
$1 + 49$
50

Day 8

① $28 + 6 \cdot 3 - 1$
$28 + 18 - 1$
$46 - 1$
45

② $9 \cdot 6 \div 2 + 4$
$54 \div 2 + 4$
$27 + 4$
31

③ $36 \div 6 + 15 \cdot 2$
$6 + 15 \cdot 2$
$6 + 30$
36

④ $11 - 5 \cdot 2 + 3$
$11 - 10 + 3$
$1 + 3$
4

⑤ $22 + 18 - 40 \div 8$
$22 + 18 - 5$
$40 - 5$
35

⑥ $35 - 5 \div 1 \cdot 7$
$35 - 5 \cdot 7$
$35 - 35$
0

⑦ $50 \div 50 \cdot 20 + 20$
$1 \cdot 20 + 20$
$20 + 20$
40

⑧ $7 \cdot 16 + 21 - 11$
$112 + 21 - 11$
$133 - 11$
122

⑨ $9 + 2 - 1 \cdot 8$
$9 + 2 - 8$
$11 - 8$
3

⑩ $13 - 5 \cdot 0 + 34$
$13 - 0 + 34$
$13 + 34$
47

Day 9

① $7 \cdot 13 - 14 \div 7$
$91 - 14 \div 7$
$91 - 2$
89

② $27 + 0 \cdot 6 - 6$
$27 + 0 - 6$
$27 - 6$
21

③ $41 - 20 \div 10 \cdot 9$
$41 - 2 \cdot 9$
$41 - 18$
23

④ $8 \div 8 + 3 \cdot 3$
$1 + 3 \cdot 3$
$1 + 9$
10

⑤ $2 \cdot 22 + 17 - 16$
$44 + 17 - 16$
$61 - 16$
45

⑥ $5 + 39 - 36 \div 9$
$5 + 39 - 4$
$44 - 4$
40

⑦ $44 - 2 \div 2 + 4$
$44 - 1 + 4$
$43 + 4$
47

⑧ $12 \div 6 \cdot 37 - 33$
$2 \cdot 37 - 33$
$74 - 33$
41

⑨ $5 + 3 - 15 \div 5$
$5 + 3 - 3$
$8 - 3$
5

⑩ $6 \cdot 10 + 23 - 5$
$60 + 23 - 5$
$83 - 5$
78

Day 10

① $36 \div 9 + 11 \cdot 3$
$4 + 11 \cdot 3$
$4 + 33$
37

② $18 - 2 \cdot 6 + 48$
$18 - 12 + 48$
$6 + 48$
54

③ $12 \cdot 1 - 6 + 2$
$12 - 6 + 2$
$6 + 2$
8

④ $4 + 14 \div 1 \cdot 8$
$4 + 14 \cdot 8$
$4 + 112$
116

⑤ $9 - 3 \cdot 3 + 6$
$9 - 9 + 6$
$0 + 6$
6

⑥ $36 \div 12 + 24 - 11$
$3 + 24 - 11$
$27 - 11$
16

⑦ $17 \cdot 3 - 19 + 25$
$51 - 19 + 25$
$32 + 25$
57

⑧ $16 + 16 \div 16 - 8$
$16 + 1 - 8$
$17 - 8$
9

⑨ $41 - 12 + 1 \cdot 0$
$41 - 12 + 0$
$29 + 0$
29

⑩ $2 \cdot 10 - 18 \div 2$
$20 - 18 \div 2$
$20 - 9$
11

Day 11

① $1 + 17 \cdot 12 - 8$
$1 + 204 - 8$
$205 - 8$
197

② $35 \div 5 - 2 \cdot 3$
$7 - 2 \cdot 3$
$7 - 6$
1

③ $6 \cdot 1 + 6 \div 2$
$6 + 6 \div 2$
$6 + 3$
9

④ $26 - 12 \div 6 + 4$
$26 - 2 + 4$
$24 + 4$
28

⑤ $43 + 8 - 16 \cdot 3$
$43 + 8 - 48$
$51 - 48$
3

⑥ $9 \div 9 \cdot 2 - 2$
$1 \cdot 2 - 2$
$2 - 2$
0

⑦ $1 \cdot 23 + 6 - 1$
$23 + 6 - 1$
$29 - 1$
28

⑧ $14 - 3 \cdot 2 + 17$
$14 - 6 + 17$
$8 + 17$
25

⑨ $50 \div 25 - 1 \cdot 1$
$2 - 1 \cdot 1$
$2 - 1$
1

⑩ $13 - 15 \div 3 + 5$
$13 - 5 + 5$
$8 + 5$
13

Day 12

① $7 \cdot 7 - 34 \div 2$
$49 - 34 \div 2$
$49 - 17$
32

② $24 - 4 \div 2 \cdot 1$
$24 - 2 \cdot 1$
$24 - 2$
22

③ $3 + 9 \cdot 9 - 3$
$3 + 81 - 3$
$84 - 3$
81

④ $45 \div 5 + 0 - 0$
$9 + 0 - 0$
$9 - 0$
9

⑤ $20 - 4 \div 2 \cdot 8$
$20 - 2 \cdot 8$
$20 - 16$
4

⑥ $11 \cdot 6 - 26 + 6$
$66 - 26 + 6$
$40 + 6$
46

⑦ $8 + 19 \cdot 21 \div 7$
$8 + 399 \div 7$
$8 + 57$
65

⑧ $14 \div 14 + 3 - 2$
$1 + 3 - 2$
$4 - 2$
2

⑨ $8 - 4 \div 1 + 1$
$8 - 4 + 1$
$4 + 1$
5

⑩ $26 \cdot 2 - 37 + 8$
$52 - 37 + 8$
$15 + 8$
23

Day 13

① 50 - 26 + 8 ÷ 4
50 - 26 + 2
24 + 2
26

② 28 + 18 ÷ 9 - 28
28 + 2 - 28
30 - 28
2

③ 36 ÷ 6 - 2 + 1
6 - 2 + 1
4 + 1
5

④ 1 · 9 + 13 - 15
9 + 13 - 15
22 - 15
7

⑤ 13 - 8 · 1 + 2
13 - 8 + 2
5 + 2
7

⑥ 41 + 6 - 8 · 3
41 + 6 - 24
47 - 24
23

⑦ 1 · 25 ÷ 5 + 1
25 ÷ 5 + 1
5 + 1
6

⑧ 33 ÷ 3 + 27 · 3
11 + 27 · 3
11 + 81
92

⑨ 41 + 16 - 16 ÷ 4
41 + 16 - 4
57 – 4
53

⑩ 19 - 7 · 2 + 7
19 - 14 + 7
5 + 7
12

Day 14

① 6 · 9 - 50 + 2
54 - 50 + 2
4 + 2
6

② 19 - 18 · 3 ÷ 3
19 - 54 ÷ 3
19 - 18
1

③ 42 + 49 ÷ 7 - 40
42 + 7 - 40
49 - 40
9

④ 38 ÷ 2 - 1 · 5
19 - 1 · 5
19 - 5
14

⑤ 8 · 7 + 16 ÷ 8
56 + 16 ÷ 8
56 + 2
58

⑥ 15 - 14 · 1 + 4
15 - 14 + 4
1 + 4
5

⑦ 10 - 4 ÷ 1 · 1
10 - 4 · 1
10 - 4
6

⑧ 17 + 6 - 30 ÷ 5
17 + 6 - 6
23 - 6
17

⑨ 11 · 2 + 25 - 22
22 + 25 - 22
47 - 22
25

⑩ 44 ÷ 4 · 13 + 7
11 · 13 + 7
143 + 7
150

Day 15

① 35 - 14 ÷ 2 · 5
35 - 7 · 5
35 - 35
0

② 12 + 37 · 24 ÷ 8
12 + 888 ÷ 8
12 + 111
123

③ 4 · 26 - 50 + 40
104 - 50 + 40
54 + 40
94

④ 9 ÷ 9 + 5 - 2
1 + 5 - 2
6 - 2
4

⑤ 0 + 1 · 1 ÷ 1
0 + 1 ÷ 1
0 + 1
1

⑥ 6 · 9 ÷ 9 + 25
54 ÷ 9 + 25
6 + 25
31

⑦ 6 · 6 - 13 + 8
36 - 13 + 8
23 + 8
31

⑧ 27 ÷ 3 + 24 · 4
9 + 24 · 4
9 + 96
105

⑨ 48 - 28 + 5 · 43
48 - 28 + 215
20 + 215
235

⑩ 15 + 5 · 7 - 7
15 + 35 - 7
50 – 7
43

Day 16

① $6 + 7^2 - 18 \div 3$
$6 + 49 - 18 \div 3$
$6 + 49 - 6$
$55 - 6$
49

② $24 + 3^2 \cdot 4 \div 2$
$24 + 9 \cdot 4 \div 2$
$24 + 36 \div 2$
$24 + 18$
42

③ $21 - 6 \cdot 3 + 8^2$
$21 - 6 \cdot 3 + 64$
$21 - 18 + 64$
$3 + 64$
67

④ $6^2 \div 3 \cdot 2 + 2^2$
$36 \div 3 \cdot 2 + 2^2$
$36 \div 3 \cdot 2 + 4$
$12 \cdot 2 + 4$
$24 + 4$
28

⑤ $31 + 4^2 \div 2 \cdot 6$
$31 + 16 \div 2 \cdot 6$
$31 + 8 \cdot 6$
$31 + 48$
79

⑥ $20 \div 4 \cdot 3^2 - 5^2$
$20 \div 4 \cdot 9 - 5^2$
$20 \div 4 \cdot 9 - 25$
$5 \cdot 9 - 25$
$45 - 25$
20

⑦ $230 - 8 \div 2^2 \cdot 7$
$230 - 8 \div 4 \cdot 7$
$230 - 2 \cdot 7$
$230 - 14$
216

⑧ $9^2 + 24 \div 8 - 2$
$81 + 24 \div 8 - 2$
$81 + 3 - 2$
$84 - 2$
82

⑨ $3 \cdot 5 + 4^2 \cdot 10$
$3 \cdot 5 + 16 \cdot 10$
$15 + 16 \cdot 10$
$15 + 160$
175

⑩ $5^2 \cdot 2 - 40 \div 5$
$25 \cdot 2 - 40 \div 5$
$50 - 40 \div 5$
$50 - 8$
42

Day 17

① $40 \div 10 + 1 \cdot 4^2$
$40 \div 10 + 1 \cdot 16$
$4 + 1 \cdot 16$
$4 + 16$
20

② $5^2 \cdot 4 \div 20 + 50$
$25 \cdot 4 \div 20 + 50$
$100 \div 20 + 50$
$5 + 50$
55

③ $31 + 6 - 9^2 \div 3$
$31 + 6 - 81 \div 3$
$31 + 6 - 27$
$37 - 27$
10

④ $49 - 2^2 \cdot 8 + 2$
$49 - 4 \cdot 8 + 2$
$49 - 32 + 2$
$17 + 2$
19

⑤
$1 \cdot 2^2 + 3^2 - 2$
$1 \cdot 4 + 3^2 - 2$
$1 \cdot 4 + 9 - 2$
$4 + 9 - 2$
$13 - 2$
11

⑥ $5^2 + 16 \div 16 - 10$
$25 + 16 \div 16 - 10$
$25 + 1 - 10$
$26 - 10$
16

⑦ $18 \div 9 - 1^2 + 1$
$18 \div 9 - 1 + 1$
$2 - 1 + 1$
$1 + 1$
2

⑧ $2^2 \cdot 2^2 - 6 + 5$
$4 \cdot 2^2 - 6 + 5$
$4 \cdot 4 - 6 + 5$
$16 - 6 + 5$
$10 + 5$
15

⑨ $7 + 6 \cdot 6 \div 3^2$
$7 + 6 \cdot 6 \div 9$
$7 + 36 \div 9$
$7 + 4$
11

⑩ $32 - 12 + 6^2 \cdot 1$
$32 - 12 + 36 \cdot 1$
$32 - 12 + 36$
$20 + 36$
56

Day 18

① $8^2 - 9 \cdot 7 + 8^2$
$64 - 9 \cdot 7 + 8^2$
$64 - 9 \cdot 7 + 64$
$64 - 63 + 64$
$1 + 64$
65

② $2^2 + 3^2 - 18 \div 6$
$4 + 3^2 - 18 \div 6$
$4 + 9 - 18 \div 6$
$4 + 9 - 3$
$13 - 3$
10

③ $2 \cdot 5^2 + 41 - 34$
$2 \cdot 25 + 41 - 34$
$50 + 41 - 34$
$91 - 34$
57

④ $16 \div 4^2 + 1 \cdot 7^2$
$16 \div 16 + 1 \cdot 7^2$
$16 \div 16 + 1 \cdot 49$
$1 + 1 \cdot 49$
$1 + 49$
50

⑤ $14 + 39 - 6^2 \div 3$
$14 + 39 - 36 \div 3$
$14 + 39 - 12$
$53 - 12$
41

⑥ $48 - 18 \div 3^2 + 1$
$48 - 18 \div 9 + 1$
$48 - 2 + 1$
$46 + 1$
47

⑦ $2 \cdot 2 + 20 - 4^2$
$2 \cdot 2 + 20 - 16$
$4 + 20 - 16$
$24 - 16$
8

⑧ $27 \div 3^2 - 1 \cdot 2$
$27 \div 9 - 1 \cdot 2$
$3 - 1 \cdot 2$
$3 - 2$
1

⑨ $25 - 10 \div 2 + 1^2$
$25 - 10 \div 2 + 1$
$25 - 5 + 1$
$20 + 1$
21

⑩ $1^2 + 5 \cdot 2^2 - 1$
$1 + 5 \cdot 2^2 - 1$
$1 + 5 \cdot 4 - 1$
$1 + 20 - 1$
$21 - 1$
20

Day 19

① $26 + 7^2 - 37 \cdot 1^2$
$26 + 49 - 37 \cdot 1^2$
$26 + 49 - 37 \cdot 1$
$26 + 49 - 37$
$75 - 37$
38

② $85 - 15 \cdot 2^2 + 2$
$85 - 15 \cdot 4 + 2$
$85 - 60 + 2$
$25 + 2$
27

③ $49 \div 7^2 + 6^2 - 22$
$49 \div 49 + 6^2 - 22$
$49 \div 49 + 36 - 22$
$1 + 36 - 22$
$37 - 22$
15

④ $5^2 \cdot 3 \div 15 - 5$
$25 \cdot 3 \div 15 - 5$
$75 \div 15 - 5$
$5 - 5$
0

⑤ $20 - 4 \cdot 5 + 4^2$
$20 - 4 \cdot 5 + 16$
$20 - 20 + 16$
$0 + 16$
16

⑥ $11 + 2^2 - 35 \div 7$
$11 + 4 - 35 \div 7$
$11 + 4 - 5$
$15 - 5$
10

⑦ $6^2 \cdot 1 \div 2 - 4^2$
$36 \cdot 1 \div 2 - 4^2$
$36 \cdot 1 \div 2 - 16$
$36 \div 2 - 16$
$18 - 16$
2

⑧ $54 - 27 + 3^2 \cdot 3$
$54 - 27 + 9 \cdot 3$
$54 - 27 + 27$
$27 + 27$
54

⑨ $14 + 1^2 - 18 \div 3^2$
$14 + 1 - 18 \div 3^2$
$14 + 1 - 18 \div 9$
$14 + 1 - 2$
$15 - 2$
13

⑩ $36 \div 12 \cdot 5^2 - 50$
$36 \div 12 \cdot 25 - 50$
$3 \cdot 25 - 50$
$75 - 50$
25

Day 20

① $50 - 7^2 \cdot 1 + 4$
$50 - 49 \cdot 1 + 4$
$50 - 49 + 4$
$1 + 4$
5

② $2^2 \cdot 7 + 5^2 - 20$
$4 \cdot 7 + 5^2 - 20$
$4 \cdot 7 + 25 - 20$
$28 + 25 - 20$
$53 - 20$
33

③ $27 + 2 - 7^2 \div 7$
$27 + 2 - 49 \div 7$
$27 + 2 - 7$
$29 - 7$
22

④ $10^2 \div 10 \cdot 1 + 5^2$
$100 \div 10 \cdot 1 + 25$
$100 \div 10 \cdot 1 + 25$
$10 \cdot 1 + 25$
$10 + 25$
35

⑤ $6 \cdot 3^2 + 13 - 12$
$6 \cdot 9 + 13 - 12$
$54 + 13 - 12$
$67 - 12$
55

⑥ $32 - 8 \div 1^2 \cdot 4$
$32 - 8 \div 1 \cdot 4$
$32 - 8 \cdot 4$
$32 - 32$
0

⑦ $2^2 + 38 - 2 \cdot 4^2$
$4 + 38 - 2 \cdot 4^2$
$4 + 38 - 2 \cdot 16$
$4 + 38 - 32$
$42 - 32$
10

⑧ $7 \cdot 8 + 6^2 \div 6^2$
$7 \cdot 8 + 36 \div 6^2$
$7 \cdot 8 + 36 \div 36$
$56 + 36 \div 36$
$56 + 1$
57

⑨ $19 - 16 \div 1^2 + 2$
$19 - 16 \div 1 + 2$
$19 - 16 + 2$
$3 + 2$
5

⑩ $18 \div 3^2 \cdot 25 - 24$
$18 \div 9 \cdot 25 - 24$
$2 \cdot 25 - 24$
$50 - 24$
26

Day 21

① $(6+10) \div 8 - 1$
$(16) \div 8 - 1$
$16 \div 8 - 1$
$2 - 1$
1

② $4 + 4 \cdot (15 - 5)$
$4 + 4 \cdot (10)$
$4 + 4 \cdot 10$
$4 + 40$
44

③ $25 - (6 + 2^2 \cdot 3)$
$25 - (6 + 4 \cdot 3)$
$25 - (6 + 12)$
$25 - (18)$
$25 - 18$
7

④ $(4^2 - 9) \cdot (8 + 2)$
$(16 - 9) \cdot (8 + 2)$
$(7) \cdot (8 + 2)$
$7 \cdot (8 + 2)$
$7 \cdot (10)$
$7 \cdot 10$
70

⑤ $(22 + 8)2 \div 6$
$(30)2 \div 6$
$30 \cdot 2 \div 6$
$60 \div 6$
10

⑥ $20 \div (2 \cdot 2) + 3$
$20 \div (4) + 3$
$20 \div 4 + 3$
$5 + 3$
8

⑦ $(4 + 20) \div (3^2 - 3)$
$(24) \div (3^2 - 3)$
$24 \div (3^2 - 3)$
$24 \div (9 - 3)$
$24 \div (6)$
$24 \div 6$
4

⑧ $14 \div (9 + 7 - 2)$
$14 \div (16 - 2)$
$14 \div (14)$
$14 \div 14$
1

⑨ $(3 + 5 \cdot 4)10$
$(3 + 20)10$
$(23)10$
$23 \cdot 10$
230

⑩ $32 \div 8(6 + 4)$
$32 \div 8(10)$
$32 \div 8 \cdot 10$
$4(10)$
$4 \cdot 10$
40

Day 22

① $6 + 6^2 \div (9 - 3)$
$6 + 6^2 \div (6)$
$6 + 6^2 \div 6$
$6 + 36 \div 6$
$6 + 6$
12

② $2^2 + 4(30 - 5^2)$
$2^2 + 4(30 - 25)$
$2^2 + 4(5)$
$2^2 + 4 \cdot 5$
$4 + 4 \cdot 5$
$4 + 20$
24

③ $100 - (3 \cdot 5 + 7^2)$
$100 - (3 \cdot 5 + 49)$
$100 - (15 + 49)$
$100 - (64)$
$100 - 64$
36

④ $(4^2 - 14)^2 (10 - 3^2)$
$(16 - 14)^2 (10 - 3^2)$
$(2)^2 (10 - 3^2)$
$2^2 (10 - 3^2)$
$2^2 (10 - 9)$
$2^2 (1)$
$2^2 \cdot 1$
$4 \cdot 1$
4

⑤ $(2 + 8)3^2 \div 3$
$(10)3^2 \div 3$
$10 \cdot 3^2 \div 3$
$10 \cdot 9 \div 3$
$90 \div 3$
30

⑥ $5^2 \div (7 - 2) + 14$
$5^2 \div (5) + 14$
$5^2 \div 5 + 14$
$25 \div 5 + 14$
$5 + 14$
19

⑦ $150 \div (8 - 3)^2 \cdot 2$
$150 \div (5)^2 \cdot 2$
$150 \div 5^2 \cdot 2$
$150 \div 25 \cdot 2$
$6 \cdot 2$
12

⑧ $18 \div (6^2 - 5 \cdot 6)$
$18 \div (36 - 5 \cdot 6)$
$18 \div (36 - 30)$
$18 \div (6)$
$18 \div 6$
3

⑨ $(8^2 + 3 \cdot 4)10$
$(64 + 3 \cdot 4)10$
$(64 + 12)10$
$(76)10$
$76 \cdot 10$
760

⑩ $(8 + 4^2) \div (5 + 3)$
$(8 + 16) \div (5 + 3)$
$(24) \div (5 + 3)$
$24 \div (8)$
$24 \div 8$
3

Day 23

① $(2 \cdot 4^2)(27 - 26)$
$(2 \cdot 16)(27 - 26)$
$(32)(27 - 26)$
$32 \cdot (1)$
$32 \cdot 1$
32

② $9(3 + 7^2) - 5$
$9(3 + 49) - 5$
$9(52) - 5$
$9 \cdot 52 - 5$
$468 - 5$
463

③ $2^2 + 6(25 \div 5^2)$
$2^2 + 6(25 \div 25)$
$2^2 + 6(1)$
$2^2 + 6 \cdot 1$
$4 + 6 \cdot 1$
$4 + 6$
10

④ $(3^2 + 7 \cdot 3^2)2$
$(9 + 7 \cdot 3^2)2$
$(9 + 7 \cdot 9)2$
$(9 + 63)2$
$(72)2$
$72 \cdot 2$
144

⑤ $(47 - 46) + 6^2 \cdot 6$
$(1) + 6^2 \cdot 6$
$1 + 6^2 \cdot 6$
$1 + 36 \cdot 6$
$1 + 216$
217

⑥ $(10^2 + 10) \div (10 + 1^2)$
$(100 + 10) \div (10 + 1^2)$
$(110) \div (10 + 1^2)$
$110 \div (10 + 1^2)$
$110 \div (10 + 1)$
$110 \div (11)$
$110 \div 11$
10

⑦ $49 + (45 \div 3^2 - 2^2)$
$49 + (45 \div 9 - 2^2)$
$49 + (45 \div 9 - 4)$
$49 + (5 - 4)$
$49 + (1)$
$49 + 1$
50

⑧ $(14 \div 7) - 1 \cdot 1^2$
$(2) - 1 \cdot 1^2$
$2 - 1 \cdot 1^2$
$2 - 1 \cdot 1$
$2 - 1$
1

⑨ $(18^2 \div 18) - (3^2 + 8)$
$(324 \div 18) - (3^2 + 8)$
$(18) - (3^2 + 8)$
$18 - (3^2 + 8)$
$18 - (9 + 8)$
$18 - (17)$
$18 - 17$
1

⑩ $35 + (9^2 - 7^2) + 6$
$35 + (81 - 7^2) + 6$
$35 + (81 - 49) + 6$
$35 + (32) + 6$
$35 + 32 + 6$
$67 + 6$
73

Day 24

① $8^2 - (9^2 \div 3) + 5^2$
$8^2 - (81 \div 3) + 5^2$
$8^2 - (27) + 5^2$
$8^2 - 27 + 5^2$
$64 - 27 + 5^2$
$64 - 27 + 25$
$37 + 25$
62

② $(11 + 1^2) \div 3 + 6$
$(11 + 1) \div 3 + 6$
$(12) \div 3 + 6$
$12 \div 3 + 6$
$4 + 6$
10

③ $32 + (4 \cdot 3^2 \div 2)$
$32 + (4 \cdot 9 \div 2)$
$32 + (36 \div 2)$
$32 + (18)$
$32 + 18$
50

④ $(42 + 7) - (1^2 \cdot 7^2)$
$(49) - (1^2 \cdot 7^2)$
$49 - (1^2 \cdot 7^2)$
$49 - (1 \cdot 7^2)$
$49 - (1 \cdot 49)$
$49 - (49)$
$49 - 49$
0

⑤ $(4^2 - 14 + 4^2) \div 6$
$(16 - 14 + 4^2) \div 6$
$(16 - 14 + 16) \div 6$
$(2 + 16) \div 6$
$(18) \div 6$
$18 \div 6$
3

⑥ $1 \cdot 5^2 \div (50 - 5^2)$
$1 \cdot 5^2 \div (50 - 25)$
$1 \cdot 5^2 \div (25)$
$1 \cdot 5^2 \div 25$
$1 \cdot 25 \div 25$
$25 \div 25$
1

⑦ $(34 - 6) \div 2^2 - 2$
$(28) \div 2^2 - 2$
$28 \div 2^2 - 2$
$28 \div 4 - 2$
$7 - 2$
5

⑧ $(10^2 \div 50) + (3 \cdot 10)$
$(100 \div 50) + (3 \cdot 10)$
$(2) + (3 \cdot 10)$
$2 + (3 \cdot 10)$
$2 + (30)$
$2 + 30$
32

⑨ $36 \div (5^2 - 7 \cdot 3)$
$36 \div (25 - 7 \cdot 3)$
$36 \div (25 - 21)$
$36 \div (4)$
$36 \div 4$
9

⑩ $(7 \cdot 4 - 11 + 12)^2$
$(28 - 11 + 12)^2$
$(17 + 12)^2$
$(29)^2$
29^2
841

Day 25

① $35 \div 7\,(50 - 7^2)$
$35 \div 7\,(50 - 49)$
$35 \div 7\,(1)$
$35 \div 7 \cdot 1$
$5 \cdot 1$
5

② $10(9^2 \div 9^2 - 1)$
$10(81 \div 9^2 - 1)$
$10(81 \div 81 - 1)$
$10(1 - 1)$
$10(0)$
$10 \cdot 0$
0

③ $(50 + 3^2 \cdot 2) \div 2^2$
$(50 + 9 \cdot 2) \div 2^2$
$(50 + 18) \div 2^2$
$(68) \div 2^2$
$68 \div 2^2$
$68 \div 4$
17

④ $8^2 \div (28 - 26) + 1^2$
$8^2 \div (2) + 1^2$
$8^2 \div 2 + 1^2$
$64 \div 2 + 1^2$
$64 \div 2 + 1$
$32 + 1$
33

⑤ $(18 - 4^2)\,(5 + 12)$
$(18 - 16)\,(5 + 12)$
$(2)\,(5 + 12)$
$2\,(5 + 12)$
$2\,(17)$
$2 \cdot 17$
34

⑥ $(50 - 6^2 + 1) - 3$
$(50 - 36 + 1) - 3$
$(14 + 1) - 3$
$(15) - 3$
$15 - 3$
12

⑦ $10(22 \div 11) + 5^2$
$10(2) + 5^2$
$10 \cdot 2 + 5^2$
$10 \cdot 2 + 25$
$20 + 25$
45

⑧ $18 - (3^2 \cdot 9 \div 9)$
$18 - (9 \cdot 9 \div 9)$
$18 - (81 \div 9)$
$18 - (9)$
$18 - 9$
9

⑨ $(6^2 \div 18 - 2) + 27$
$(36 \div 18 - 2) + 27$
$(2 - 2) + 27$
$(0) + 27$
$0 + 27$
27

⑩ $(50 \cdot 4) - 11 + 9^2$
$(200) - 11 + 9^2$
$200 - 11 + 9^2$
$200 - 11 + 81$
$189 + 81$
270

Day 26

① $7^2 \cdot 1 - (22 + 9)$
$7^2 \cdot 1 - (31)$
$7^2 \cdot 1 - 31$
$49 \cdot 1 - 31$
$49 - 31$
18

② $16 \div 8(3 \cdot 2^2)$
$16 \div 8(3 \cdot 4)$
$16 \div 8(12)$
$16 \div 8 \cdot 12$
$2 \cdot 12$
24

③ $28 - (4 \cdot 4 \div 4)^2$
$28 - (16 \div 4)^2$
$28 - (4)^2$
$28 - 4^2$
$28 - 16$
12

④ $(6 + 5)^2 - (1 + 6)^2$
$(11)^2 - (1 + 6)^2$
$11^2 - (1 + 6)^2$
$11^2 - (7)^2$
$11^2 - 7^2$
$121 - 7^2$
$121 - 49$
72

⑤ $(9^2 \cdot 1^2 \div 3^2) + 10$
$(81 \cdot 1^2 \div 3^2) + 10$
$(81 \cdot 1 \div 3^2) + 10$
$(81 \cdot 1 \div 9) + 10$
$(81 \div 9) + 10$
$(9) + 10$
$9 + 10$
19

⑥ $(38 - 28) \cdot 6^2 \div 36$
$(10) \cdot 6^2 \div 36$
$10 \cdot 6^2 \div 36$
$10 \cdot 36 \div 36$
$360 \div 36$
10

⑦ $296 - (21 + 10 \cdot 5^2)$
$296 - (21 + 10 \cdot 25)$
$296 - (21 + 250)$
$296 - (271)$
$296 - 271$
25

⑧ $12 + (6^2 \div 2^2 - 9)$
$12 + (36 \div 2^2 - 9)$
$12 + (36 \div 4 - 9)$
$12 + (9 - 9)$
$12 + (0)$
$12 + 0$
12

⑨ $7^2 - (8 + 7) \cdot 3$
$7^2 - (15) \cdot 3$
$7^2 - 15 \cdot 3$
$49 - 15 \cdot 3$
$49 - 45$
4

⑩ $(2 \cdot 7)\,(18 - 17)$
$(14)\,(18 - 17)$
$14\,(18 - 17)$
$14\,(1)$
$14 \cdot 1$
14

Day 27

① $(16 - 10)^2 + (13 + 20)$
$(6)^2 + (13 + 20)$
$6^2 + (13 + 20)$
$6^2 + (33)$
$6^2 + 33$
$36 + 33$
69

② $2^2 \div (1 + 1 \cdot 1)$
$2^2 \div (1 + 1)$
$2^2 \div (2)$
$2^2 \div 2$
$4 \div 2$
2

③ $5 + (18 \div 3)^2 \cdot 3$
$5 + (6)^2 \cdot 3$
$5 + 6^2 \cdot 3$
$5 + 36 \cdot 3$
$5 + 108$
113

④ $(16 \cdot 5^2 - 8) + 70$
$(16 \cdot 25 - 8) + 70$
$(400 - 8) + 70$
$(392) + 70$
$392 + 70$
462

⑤ $(11 + 12)^2 - (12 \div 12)^2$
$(23)^2 - (12 \div 12)^2$
$23^2 - (12 \div 12)^2$
$23^2 - (1)^2$
$23^2 - 1^2$
$529 - 1^2$
$529 - 1$
528

⑥ $37 - (20 \cdot 4) \div 8$
$37 - (80) \div 8$
$37 - 80 \div 8$
$37 - 10$
27

⑦ $10^2 \div (5 + 3 \cdot 5)$
$10^2 \div (5 + 15)$
$10^2 \div (20)$
$10^2 \div 20$
$100 \div 20$
5

⑧ $27 + 7(2 \cdot 1)$
$27 + 7(2)$
$27 + 7 \cdot 2$
$27 + 14$
41

⑨ $30 + (39 - 9) \div 10$
$30 + (30) \div 10$
$30 + 30 \div 10$
$30 + 3$
33

⑩ $(15 \cdot 6 - 40) + 7$
$(90 - 40) + 7$
$(50) + 7$
$50 + 7$
57

Day 28

① $29 + 36 - (10^2 \div 50)$
$29 + 36 - (100 \div 50)$
$29 + 36 - (2)$
$29 + 36 - 2$
$65 - 2$
63

② $(3^2 - 6) - (1 + 1^2)$
$(9 - 6) - (1 + 1^2)$
$(3) - (1 + 1^2)$
$3 - (1 + 1^2)$
$3 - (1 + 1)$
$3 - 2$
1

③ $(4 \cdot 2^2 + 3) - 4^2$
$(4 \cdot 4 + 3) - 4^2$
$(16 + 3) - 4^2$
$(19) - 4^2$
$19 - 4^2$
$19 - 16$
3

④ $(81 \div 9 + 6^2) \cdot 1$
$(81 \div 9 + 36) \cdot 1$
$(9 + 36) \cdot 1$
$(45) \cdot 1$
$45 \cdot 1$
45

⑤ $(7^2 - 49) + (2 \cdot 1)^2$
$(49 - 49) + (2 \cdot 1)^2$
$(0) + (2 \cdot 1)^2$
$0 + (2 \cdot 1)^2$
$0 + (2)^2$
$0 + 2^2$
$0 + 4$
4

⑥ $(2 + 8)(8 + 9)$
$(10)(8 + 9)$
$10 \cdot (8 + 9)$
$10 \cdot (17)$
$10 \cdot 17$
170

⑦ $10^2 \cdot (5 - 10 \div 5)$
$10^2 \cdot (5 - 2)$
$10^2 \cdot (3)$
$10^2 \cdot 3$
$100 \cdot 3$
300

⑧ $(4 + 3 \cdot 8) \div 7$
$(4 + 24) \div 7$
$(28) \div 7$
$28 \div 7$
4

⑨ $6^2 + (6^2 - 4 \div 4)$
$6^2 + (36 - 4 \div 4)$
$6^2 + (36 - 1)$
$6^2 + (35)$
$6^2 + 35$
$36 + 35$
71

⑩ $(2^2)(81 \div 3^2) - 20$
$(4)(81 \div 3^2) - 20$
$4(81 \div 3^2) - 20$
$4(81 \div 9) - 20$
$4(9) - 20$
$4 \cdot 9 - 20$
$36 - 20$
16

Day 29

① $3^2 - (1 \cdot 2^2 + 1^2)$
$3^2 - (1 \cdot 4 + 1^2)$
$3^2 - (1 \cdot 4 + 1)$
$3^2 - (4 + 1)$
$3^2 - (5)$
$3^2 - 5$
$9 - 5$
4

② $5 \cdot 9 + (12 - 2^2)$
$5 \cdot 9 + (12 - 4)$
$5 \cdot 9 + (8)$
$5 \cdot 9 + 8$
$45 + 8$
53

③ $(18 + 8 - 19)^2 - 12$
$(26 - 19)^2 - 12$
$(7)^2 - 12$
$7^2 - 12$
$49 - 12$
37

④ $(5 - 1)^2 \div (9 - 1)$
$(4)^2 \div (9 - 1)$
$4^2 \div (9 - 1)$
$4^2 \div (8)$
$4^2 \div 8$
$16 \div 8$
2

⑤ $11 + (5^2 \div 5^2) - 11$
$11 + (25 \div 5^2) - 11$
$11 + (25 \div 25) - 11$
$11 + (1) - 11$
$11 + 1 - 11$
$12 - 11$
1

⑥ $49 \div 7 \cdot (6^2 - 5^2)$
$49 \div 7 \cdot (36 - 5^2)$
$49 \div 7 \cdot (36 - 25)$
$49 \div 7 \cdot (11)$
$49 \div 7 \cdot 11$
$7 \cdot 11$
77

⑦ $5(2 + 2) \cdot 3^2$
$5(4) \cdot 3^2$
$5 \cdot 4 \cdot 3^2$
$5 \cdot 4 \cdot 9$
$20 \cdot 9$
180

⑧ $16 + (7^2 \cdot 1^2 - 9)$
$16 + (49 \cdot 1^2 - 9)$
$16 + (49 \cdot 1 - 9)$
$16 + (49 - 9)$
$16 + (40)$
$16 + 40$
56

⑨ $(9^2 + 3^2 - 50) \div 8$
$(81 + 3^2 - 50) \div 8$
$(81 + 9 - 50) \div 8$
$(90 - 50) \div 8$
$(40) \div 8$
$40 \div 8$
5

⑩ $(9 - 2)^2 - (9 \cdot 2)$
$(7)^2 - (9 \cdot 2)$
$7^2 - (9 \cdot 2)$
$7^2 - (18)$
$7^2 - 18$
$49 - 18$
31

Day 30

① $20 \cdot 6^2 - (8 + 1)^2$
$20 \cdot 6^2 - (9)^2$
$20 \cdot 6^2 - 9^2$
$20 \cdot 36 - 9^2$
$20 \cdot 36 - 81$
$720 - 81$
639

② $160 \div (1 + 3 \cdot 5)$
$160 \div (1 + 15)$
$160 \div (16)$
$160 \div 16$
10

③ $9 + (3 \cdot 3 \div 3)^2$
$9 + (9 \div 3)^2$
$9 + (3)^2$
$9 + 3^2$
$9 + 9$
18

④ $(35 + 4^2) - (1 + 2^2)^2$
$(35 + 16) - (1 + 2^2)^2$
$(51) - (1 + 2^2)^2$
$51 - (1 + 2^2)^2$
$51 - (1 + 4)^2$
$51 - (5)^2$
$51 - 5^2$
$51 - 25$
26

⑤ $(49 - 7^2 + 3) \div 3$
$(49 - 49 + 3) \div 3$
$(0 + 3) \div 3$
$(3) \div 3$
$3 \div 3$
1

⑥ $(36 \div 6) \cdot 4 - 2^2$
$(6) \cdot 4 - 2^2$
$6 \cdot 4 - 2^2$
$6 \cdot 4 - 4$
$24 - 4$
20

⑦ $19 + (8^2 - 6 \cdot 7)$
$19 + (64 - 6 \cdot 7)$
$19 + (64 - 42)$
$19 + (22)$
$19 + 22$
41

⑧ $5^2 + (6 \div 3) - 5^2$
$5^2 + (2) - 5^2$
$5^2 + 2 - 5^2$
$25 + 2 - 5^2$
$25 + 2 - 25$
$27 - 25$
2

⑨ $(5^2 - 24 + 7) \div 4$
$(25 - 24 + 7) \div 4$
$(1 + 7) \div 4$
$(8) \div 4$
$8 \div 4$
2

⑩ $(1 \cdot 6)^2 - (4 - 2)^2$
$(6)^2 - (4 - 2)^2$
$6^2 - (4 - 2)^2$
$6^2 - (2)^2$
$6^2 - 2^2$
$36 - 2^2$
$36 - 4$
32

Day 31

① $(349 - 7^2)(1 + 2)$
$(349 - 49)(1 + 2)$
$(300)(1 + 2)$
$300(1 + 2)$
$300(3)$
$300 \cdot 3$
900

② $(3 + 2)^2 \div (30 - 5^2)$
$(5)^2 \div (30 - 5^2)$
$5^2 \div (30 - 5^2)$
$5^2 \div (30 - 25)$
$5^2 \div (5)$
$5^2 \div 5$
$25 \div 5$
5

③ $(6^2 - 1^2 + 5^2) \div 30$
$(36 - 1^2 + 5^2) \div 30$
$(36 - 1 + 5^2) \div 30$
$(36 - 1 + 25) \div 30$
$(35 + 25) \div 30$
$(60) \div 30$
$60 \div 30$
2

④ $(9 \div 3^2 + 3^2) \cdot 4$
$(9 \div 9 + 3^2) \cdot 4$
$(9 \div 9 + 9) \cdot 4$
$(1 + 9) \cdot 4$
$(10) \cdot 4$
$10 \cdot 4$
40

⑤ $(10 - 8)^2 + (2 - 1)^2$
$(2)^2 + (2 - 1)^2$
$2^2 + (2 - 1)^2$
$2^2 + (1)^2$
$2^2 + 1^2$
$4 + 1^2$
$4 + 1$
5

⑥ $(6^2 \div 3) - (4 \cdot 1^2)$
$(36 \div 3) - (4 \cdot 1^2)$
$(12) - (4 \cdot 1^2)$
$12 - (4 \cdot 1^2)$
$12 - (4 \cdot 1)$
$12 - (4)$
$12 - 4$
8

⑦ $45 + (6 \cdot 2^2 + 2^2)$
$45 + (6 \cdot 4 + 2^2)$
$45 + (6 \cdot 4 + 4)$
$45 + (24 + 4)$
$45 + (28)$
$45 + 28$
73

⑧ $50 \div (36 - 26)(7)$
$50 \div (10)(7)$
$50 \div 10(7)$
$50 \div 10 \cdot 7$
$5 \cdot 7$
35

⑨ $7 + (43 - 14 \div 2)$
$7 + (43 - 7)$
$7 + (36)$
$7 + 36$
43

⑩ $100(81 \div 9^2 - 1^2)$
$100(81 \div 81 - 1^2)$
$100(81 \div 81 - 1)$
$100(1 - 1)$
$100(0)$
$100 \cdot 0$
0

Day 32

① $(6 + 4)^2 - (5 \div 1)^2$
$(10)^2 - (5 \div 1)^2$
$10^2 - (5 \div 1)^2$
$10^2 - (5)^2$
$10^2 - 5^2$
$100 - 5^2$
$100 - 25$
75

② $28 - (49 \div 7^2 \cdot 28)$
$28 - (49 \div 49 \cdot 28)$
$28 - (1 \cdot 28)$
$28 - (28)$
$28 - 28$
0

③ $3 \div (2 - 1)^2 \cdot 3^2$
$3 \div (1)^2 \cdot 3^2$
$3 \div 1^2 \cdot 3^2$
$3 \div 1 \cdot 3^2$
$3 \div 1 \cdot 9$
$3 \cdot 9$
27

④ $(4 \cdot 2 - 2)^2 + 2$
$(8 - 2)^2 + 2$
$(6)^2 + 2$
$6^2 + 2$
$36 + 2$
38

⑤ $(5 \cdot 6 - 2) \div (7 - 5)$
$(30 - 2) \div (7 - 5)$
$(28) \div (7 - 5)$
$28 \div (7 - 5)$
$28 \div (2)$
$28 \div 2$
14

⑥ $6 + 12 \div (10 - 7)$
$6 + 12 \div (3)$
$6 + 12 \div 3$
$6 + 4$
10

⑦ $(4)(7^2 + 1^2 \cdot 3)$
$4(7^2 + 1^2 \cdot 3)$
$4(49 + 1^2 \cdot 3)$
$4(49 + 1 \cdot 3)$
$4(49 + 3)$
$4(52)$
$4 \cdot 52$
208

⑧ $27 - (77 \div 11)(2)$
$27 - (7)(2)$
$27 - 7(2)$
$27 - 7 \cdot 2$
$27 - 14$
13

⑨ $80 \div (1 + 3) + 10$
$80 \div (4) + 10$
$80 \div 4 + 10$
$20 + 10$
30

⑩ $(7 \cdot 2 - 2) \div 6$
$(14 - 2) \div 6$
$(12) \div 6$
$12 \div 6$
2

Day 33

① $18 \div 9 + (4 - 3)^2$
$18 \div 9 + (1)^2$
$18 \div 9 + 1^2$
$18 \div 9 + 1$
$2 + 1$
3

② $3(1 + 7 \cdot 2)$
$3(1 + 14)$
$3(15)$
$3 \cdot 15$
45

③ $6^2 + (4^2 - 2 \cdot 3)$
$6^2 + (16 - 2 \cdot 3)$
$6^2 + (16 - 6)$
$6^2 + (10)$
$6^2 + 10$
$36 + 10$
46

④ $(8 + 0)^2 - (24 + 27)$
$(8)^2 - (24 + 27)$
$8^2 - (24 + 27)$
$8^2 - (51)$
$8^2 - 51$
$64 - 51$
13

⑤ $(35 + 25 - 12) \div 8$
$(60 - 12) \div 8$
$(48) \div 8$
$48 \div 8$
6

⑥ $150 \div 15 \cdot (4^2 - 2^2)$
$150 \div 15 \cdot (16 - 2^2)$
$150 \div 15 \cdot (16 - 4)$
$150 \div 15 \cdot (12)$
$150 \div 15 \cdot 12$
$10 \cdot 12$
120

⑦ $19 - (3 \div 3 \cdot 1)^2$
$19 - (1 \cdot 1)^2$
$19 - (1)^2$
$19 - 1$
18

⑧ $5 + (6 \div 6) - 2^2$
$5 + (1) - 2^2$
$5 + 1 - 2^2$
$5 + 1 - 4$
$6 - 4$
2

⑨ $(3 + 15 \div 3)^2 + 16$
$(3 + 5)^2 + 16$
$(8)^2 + 16$
$8^2 + 16$
$64 + 16$
80

⑩ $(7^2 \cdot 4^2) + (30 - 21)$
$(49 \cdot 4^2) + (30 - 21)$
$(49 \cdot 16) + (30 - 21)$
$(784) + (30 - 21)$
$784 + (30 - 21)$
$784 + (9)$
$784 + 9$
793

Day 34

① $(8 - 8) \cdot (3^2 + 5^2)$
$(0) \cdot (3^2 + 5^2)$
$0 \cdot (3^2 + 5^2)$
$0 \cdot (9 + 5^2)$
$0 \cdot (9 + 25)$
$0 \cdot (34)$
$0 \cdot 34$
0

② $28 \div (10 - 3 \cdot 2)$
$28 \div (10 - 6)$
$28 \div (4)$
$28 \div 4$
7

③ $6 + (9 - 4) \cdot 7$
$6 + (5) \cdot 7$
$6 + 5 \cdot 7$
$6 + 35$
41

④ $(28 - 21 + 1) \cdot 3^2$
$(7 + 1) \cdot 3^2$
$(8) \cdot 3^2$
$8 \cdot 3^2$
$8 \cdot 9$
72

⑤ $(8^2 + 1^2) - (12 \div 3)$
$(64 + 1^2) - (12 \div 3)$
$(64 + 1) - (12 \div 3)$
$(65) - (12 \div 3)$
$65 - (12 \div 3)$
$65 - (4)$
$65 - 4$
61

⑥ $12 - (6^2 \div 3) \div 2$
$12 - (36 \div 3) \div 2$
$12 - (12) \div 2$
$12 - 12 \div 2$
$12 - 6$
6

⑦ $30 + (5 - 3 \div 1)^2$
$30 + (5 - 3)^2$
$30 + (2)^2$
$30 + 2^2$
$30 + 4$
34

⑧ $2 + 5(2 \cdot 1)$
$2 + 5(2)$
$2 + 5 \cdot 2$
$2 + 10$
12

⑨ $6^2 + (5^2 - 7) \div 6$
$6^2 + (25 - 7) \div 6$
$6^2 + (18) \div 6$
$6^2 + 18 \div 6$
$36 + 18 \div 6$
$36 + 3$
39

⑩ $(30 - 10 \cdot 2) + 7$
$(30 - 20) + 7$
$(10) + 7$
$10 + 7$
17

Day 35

① $12 \div (2 \cdot 3 - 4)^2$
$12 \div (6 - 4)^2$
$12 \div (2)^2$
$12 \div 2^2$
$12 \div 4$
3

② $30 - 4^2 \div (14 - 6)$
$30 - 4^2 \div (8)$
$30 - 4^2 \div 8$
$30 - 16 \div 8$
$30 - 2$
28

③ $(9^2 + 23 - 49) \div 5$
$(81 + 23 - 49) \div 5$
$(104 - 49) \div 5$
$(55) \div 5$
$55 \div 5$
11

④ $(7 + 2)\ (1 + 4)$
$(9)\ (1 + 4)$
$9\ (1 + 4)$
$9\ (5)$
$9 \cdot 5$
45

⑤ $8^2 + (81 \div 9) - 7^2$
$8^2 + (9) - 7^2$
$8^2 + 9 - 7^2$
$64 + 9 - 7^2$
$64 + 9 - 49$
$73 - 49$
24

⑥ $60 \div 3 \cdot (28 \div 7)$
$60 \div 3 \cdot (4)$
$60 \div 3 \cdot 4$
$20 \cdot 4$
80

⑦ $8 \div (6 \div 3)^2 \cdot 5$
$8 \div (2)^2 \cdot 5$
$8 \div 2^2 \cdot 5$
$8 \div 4 \cdot 5$
$2 \cdot 5$
10

⑧ $16 + (3 + 2^2 \cdot 9)$
$16 + (3 + 4 \cdot 9)$
$16 + (3 + 36)$
$16 + (39)$
$16 + 39$
55

⑨ $(9^2 + 3^2 - 50) \div 2$
$(81 + 3^2 - 50) \div 2$
$(81 + 9 - 50) \div 2$
$(90 - 50) \div 2$
$(40) \div 2$
$40 \div 2$
20

⑩ $(6^2 - 24) - (3^2 \cdot 1)$
$(36 - 24) - (3^2 \cdot 1)$
$(12) - (3^2 \cdot 1)$
$12 - (3^2 \cdot 1)$
$12 - (9 \cdot 1)$
$12 - (9)$
$12 - 9$
3

Day 36

① $((6 + 7^2) \div 5)2$
$((6 + 49) \div 5)2$
$((55) \div 5)2$
$(55 \div 5)2$
$(11)2$
$11 \cdot 2$
22

② $7((40 - 6^2) \div 2)$
$7((40 - 36) \div 2)$
$7((4) \div 2)$
$7(4 \div 2)$
$7(2)$
$7 \cdot 2$
14

③ $4(72 - (4 \cdot 7 + 6^2))$
$4(72 - (4 \cdot 7 + 36))$
$4(72 - (28 + 36))$
$4(72 - (64))$
$4(72 - 64)$
$4(8)$
$4 \cdot 8$
32

④ $(8^2 - 59)((3 + 3^2) \div 2)$
$(64 - 59)((3 + 3^2) \div 2)$
$(5)((3 + 3^2) \div 2)$
$5((3 + 3^2) \div 2)$
$5((3 + 9) \div 2)$
$5((12) \div 2)$
$5(12 \div 2)$
$5(6)$
$5 \cdot 6$
30

⑤ $(4 + (6 - 4)) \div 3$
$(4 + (2)) \div 3$
$(4 + 2) \div 3$
$(6) \div 3$
$6 \div 3$
2

⑥ $2^2((7 - 2)2)$
$2^2((5)2)$
$2^2(5 \cdot 2)$
$2^2(10)$
$2^2 \cdot 10$
$4 \cdot 10$
40

⑦ $36 \div ((3 + 3)^2 \div 4)$
$36 \div ((6)^2 \div 4)$
$36 \div (6^2 \div 4)$
$36 \div (36 \div 4)$
$36 \div (9)$
$36 \div 9$
4

⑧ $56 \div (2^2(5 - 3))$
$56 \div (2^2(2))$
$56 \div (2^2 \cdot 2)$
$56 \div (4 \cdot 2)$
$56 \div (8)$
$56 \div 8$
7

⑨ $(61 - (7^2 + 6))10$
$(61 - (49 + 6))10$
$(61 - (55))10$
$(61 - 55)10$
$(6)10$
$6 \cdot 10$
60

⑩ $(12 - 4) \div ((6 + 2) \div 4)$
$(8) \div ((6 + 2) \div 4)$
$8 \div ((6 + 2) \div 4)$
$8 \div ((8) \div 4)$
$8 \div (8 \div 4)$
$8 \div (2)$
$8 \div 2$
4

Day 37

① $((31 + 23)) - (9 \div 3)^2$
$((54)) - (9 \div 3)^2$
$(54) - (9 \div 3)^2$
$54 - (9 \div 3)^2$
$54 - (3)^2$
$54 - 3^2$
$54 - 9$
45

② $80 - ((6 + 2)^2 + 2)$
$80 - ((8)^2 + 2)$
$80 - (8^2 + 2)$
$80 - (64 + 2)$
$80 - (66)$
$80 - 66$
14

③ $34 - (5^2 + 4^2 \div (1 + 1))$
$34 - (5^2 + 4^2 \div (2))$
$34 - (5^2 + 4^2 \div 2)$
$34 - (25 + 4^2 \div 2)$
$34 - (25 + 16 \div 2)$
$34 - (25 + 8)$
$34 - (33)$
$34 - 33$
1

④ $(3^2 - 2) + (7 \div (4 - 3))^2$
$(9 - 2) + (7 \div (4 - 3))^2$
$(7) + (7 \div (4 - 3))^2$
$7 + (7 \div (4 - 3))^2$
$7 + (7 \div (1))^2$
$7 + (7 \div 1)^2$
$7 + (7)^2$
$7 + 7^2$
$7 + 49$
56

⑤ $((4 \cdot 2 \div 2)^2 - 4) + 4^2$
$((8 \div 2)^2 - 4) + 4^2$
$((4)^2 - 4) + 4^2$
$(4^2 - 4) + 4^2$
$(16 - 4) + 4^2$
$(12) + 4^2$
$12 + 4^2$
$12 + 16$
28

⑥ $2((34 - 25) + 9^2 - 8^2)$
$2((9) + 9^2 - 8^2)$
$2(9 + 9^2 - 8^2)$
$2(9 + 81 - 8^2)$
$2(9 + 81 - 64)$
$2(90 - 64)$
$2(26)$
$2 \cdot 26$
52

⑦ $45 - (3 + (6^2 \div 9)^2)$
$45 - (3 + (36 \div 9)^2)$
$45 - (3 + (4)^2)$
$45 - (3 + 4^2)$
$45 - (3 + 16)$
$45 - (19)$
$45 - 19$
26

⑧ $14 + (6^2 + (8 \div 2 + 33))$
$14 + (6^2 + (4 + 33))$
$14 + (6^2 + (37))$
$14 + (6^2 + 37)$
$14 + (36 + 37)$
$14 + (73)$
$14 + 73$
87

⑨ $150 + ((8 - 2)^2 - 3^2)$
$150 + ((6)^2 - 3^2)$
$150 + (6^2 - 3^2)$
$150 + (36 - 3^2)$
$150 + (36 - 9)$
$150 + (27)$
$150 + 27$
177

⑩ $(4 + (24 - 6)) \div 2$
$(4 + (18)) \div 2$
$(4 + 18) \div 2$
$(22) \div 2$
$22 \div 2$
11

Day 38

① $42 - ((8 - 5)^2 + 3)$
$42 - ((3)^2 + 3)$
$42 - (3^2 + 3)$
$42 - (9 + 3)$
$42 - (12)$
$42 - 12$
30

② $120 \div ((20 - 2) + 12) - 4$
$120 \div ((18) + 12) - 4$
$120 \div (18 + 12) - 4$
$120 \div (30) - 4$
$120 \div 30 - 4$
$4 - 4$
0

③ $(16 + (5 + 3^2))2$
$(16 + (5 + 9))2$
$(16 + (14))2$
$(16 + 14)2$
$(30)2$
$30 \cdot 2$
60

④ $36 - ((6 - 3)^2 + (36 - 9))$
$36 - ((3)^2 + (36 - 9))$
$36 - (3^2 + (36 - 9))$
$36 - (3^2 + (27))$
$36 - (3^2 + 27)$
$36 - (9 + 27)$
$36 - (36)$
$36 - 36$
0

⑤ $((7^2 + 10 - 2) + 6) \div 9$
$((49 + 10 - 2) + 6) \div 9$
$((59 - 2) + 6) \div 9$
$((57) + 6) \div 9$
$(57 + 6) \div 9$
$(63) \div 9$
$63 \div 9$
7

⑥ $3((9 + 1) \cdot 2^2 - 5^2)$
$3((10) \cdot 2^2 - 5^2)$
$3(10 \cdot 2^2 - 5^2)$
$3(10 \cdot 4 - 5^2)$
$3(10 \cdot 4 - 25)$
$3(40 - 25)$
$3(15)$
$3 \cdot 15$
45

⑦ $4^2 \div ((2 \cdot 2) + 2^2)$
$4^2 \div ((4) + 2^2)$
$4^2 \div (4 + 2^2)$
$4^2 \div (4 + 4)$
$4^2 \div (8)$
$4^2 \div 8$
$16 \div 8$
2

⑧ $5((6^2 + 8 \div 2) - 33)$
$5((36 + 8 \div 2) - 33)$
$5((36 + 4) - 33)$
$5((40) - 33)$
$5(40 - 33)$
$5(7)$
$5 \cdot 7$
35

⑨ $42 \div ((7 - 2)^2 - 19)$
$42 \div ((5)^2 - 19)$
$42 \div (5^2 - 19)$
$42 \div (25 - 19)$
$42 \div (6)$
$42 \div 6$
7

⑩ $(5 + (5 - 1)^2) \div (4 + 3)$
$(5 + (4)^2) \div (4 + 3)$
$(5 + 4^2) \div (4 + 3)$
$(5 + 16) \div (4 + 3)$
$(21) \div (4 + 3)$
$21 \div (4 + 3)$
$21 \div (7)$
$21 \div 7$
3

Day 39

(1)
$2(42 + 56 \div (9 - 2))$
$2(42 + 56 \div (7))$
$2(42 + 56 \div 7)$
$2(42 + 8)$
$2(50)$
$2 \cdot 50$
100

(2)
$3((7 + 3) \cdot 5 - 7^2)$
$3((10) \cdot 5 - 7^2)$
$3(10 \cdot 5 - 7^2)$
$3(10 \cdot 5 - 49)$
$3(50 - 49)$
$3(1)$
$3 \cdot 1$
3

(3)
$30 \div ((8 - 5)^2 - 4)$
$30 \div ((3)^2 - 4)$
$30 \div (3^2 - 4)$
$30 \div (9 - 4)$
$30 \div (5)$
$30 \div 5$
6

(4)
$(100 - (5 + 3)^2) \div 9$
$(100 - (8)^2) \div 9$
$(100 - 8^2) \div 9$
$(100 - 64) \div 9$
$(36) \div 9$
$36 \div 9$
4

(5)
$10((9 - 3 \div 3) - 1)$
$10((9 - 1) - 1)$
$10((8) - 1)$
$10(8 - 1)$
$10(7)$
$10 \cdot 7$
70

(6)
$9((5 - (1 + 1)^2 - 1))$
$9((5 - (2)^2 - 1))$
$9((5 - 2^2 - 1))$
$9((5 - 4 - 1))$
$9((1 - 1))$
$9((0))$
$9(0)$
$9 \cdot 0$
0

(7)
$12 + (3(2 + 1))^2$
$12 + (3(3))^2$
$12 + (3 \cdot 3)^2$
$12 + (9)^2$
$12 + 9^2$
$12 + 81$
93

(8)
$((2 + 4)^2 + 8) \div (7 - 3)$
$((6)^2 + 8) \div (7 - 3)$
$(6^2 + 8) \div (7 - 3)$
$(36 + 8) \div (7 - 3)$
$(44) \div (7 - 3)$
$44 \div (7 - 3)$
$44 \div (4)$
$44 \div 4$
11

(9)
$6 \div (100 \div (7^2 + 1))$
$6 \div (100 \div (49 + 1))$
$6 \div (100 \div (50))$
$6 \div (100 \div 50)$
$6 \div (2)$
$6 \div 2$
3

(10)
$3(3^2 + 4(3 + 1))$
$3(3^2 + 4(4))$
$3(3^2 + 4 \cdot 4)$
$3(9 + 4 \cdot 4)$
$3(9 + 16)$
$3 \cdot (25)$
$3 \cdot 25$
75

Day 40

(1)
$(32 - 5^2)((20 + 28) \div 8)$
$(32 - 25)((20 + 28) \div 8)$
$(7)((20 + 28) \div 8)$
$7((20 + 28) \div 8)$
$7((48) \div 8)$
$7(48 \div 8)$
$7(6)$
$7 \cdot 6$
42

(2)
$(1 + 7(13 - 8)) \div 3$
$(1 + 7(5)) \div 3$
$(1 + 7 \cdot 5) \div 3$
$(1 + 35) \div 3$
$(36) \div 3$
$36 \div 3$
12

(3)
$5^2 + (4 \div (3 - 2))^2$
$5^2 + (4 \div (1))^2$
$5^2 + (4 \div 1)^2$
$5^2 + (4)^2$
$5^2 + 4^2$
$25 + 4^2$
$25 + 16$
41

(4)
$((12 - 4 \div 2) - 2)3$
$((12 - 2) - 2)3$
$((10) - 2)3$
$(10 - 2)3$
$(8)3$
$8 \cdot 3$
24

(5)
$((8 - 5)2)^2 - 32$
$((3)2)^2 - 32$
$(3 \cdot 2)^2 - 32$
$(6)^2 - 32$
$6^2 - 32$
$36 - 32$
4

(6)
$28 \div ((11 - 9) + (2^2 + 1))$
$28 \div ((2) + (2^2 + 1))$
$28 \div (2 + (2^2 + 1))$
$28 \div (2 + (4 + 1))$
$28 \div (2 + (5))$
$28 \div (2 + 5)$
$28 \div (7)$
$28 \div 7$
4

(7)
$(7^2 - 4) \div ((8 + 7) \div 3)$
$(49 - 4) \div ((8 + 7) \div 3)$
$(45) \div ((8 + 7) \div 3)$
$45 \div ((8 + 7) \div 3)$
$45 \div ((15) \div 3)$
$45 \div (15 \div 3)$
$45 \div (5)$
$45 \div 5$
9

(8)
$6(9^2 - (8 \cdot 7 + 4^2))$
$6(9^2 - (8 \cdot 7 + 16))$
$6(9^2 - (56 + 16))$
$6(9^2 - (72))$
$6(9^2 - 72)$
$6(81 - 72)$
$6(9)$
$6 \cdot 9$
54

(9)
$9 + (8^2 - 2(10 + 15))$
$9 + (8^2 - 2(25))$
$9 + (8^2 - 2 \cdot 25)$
$9 + (64 - 2 \cdot 25)$
$9 + (64 - 50)$
$9 + (14)$
$9 + 14$
23

(10)
$(50 - (5 + 6 \cdot 2))10$
$(50 - (5 + 12))10$
$(50 - (17))10$
$(50 - 17)10$
$(33)10$
$33 \cdot 10$
330

Day 41

(1)
$30 \div (2 + (8 - 4)^2 \div 2)$
$30 \div (2 + (4)^2 \div 2)$
$30 \div (2 + 4^2 \div 2)$
$30 \div (2 + 16 \div 2)$
$30 \div (2 + 8)$
$30 \div (10)$
$30 \div 10$
3

(2)
$3^2 + (3^2 - (8 - 5))$
$3^2 + (3^2 - (3))$
$3^2 + (3^2 - 3)$
$3^2 + (9 - 3)$
$3^2 + (6)$
$3^2 + 6$
$9 + 6$
15

(3)
$(26 + 16) \cdot 1^2 \div (1 \cdot 1)^2$
$(42) \cdot 1^2 \div (1 \cdot 1)^2$
$42 \cdot 1^2 \div (1 \cdot 1)^2$
$42 \cdot 1^2 \div (1)^2$
$42 \cdot 1^2 \div 1^2$
$42 \cdot 1 \div 1^2$
$42 \cdot 1 \div 1$
$42 \div 1$
42

(4)
$((6 \div 2) - 3) + 1^2 - 1^2$
$((3) - 3) + 1^2 - 1^2$
$(3 - 3) + 1^2 - 1^2$
$(0) + 1^2 - 1^2$
$0 + 1^2 - 1^2$
$0 + 1 - 1^2$
$0 + 1 - 1$
$1 - 1$
0

(5)
$(8 - (3 - 2))^2 + 37$
$(8 - (1))^2 + 37$
$(8 - 1)^2 + 37$
$(7)^2 + 37$
$7^2 + 37$
$49 + 37$
86

(6)
$120 \div ((20 - 2) + 12) - 1$
$120 \div ((18) + 12) - 1$
$120 \div (18 + 12) - 1$
$120 \div (30) - 1$
$120 \div 30 - 1$
$4 - 1$
3

(7)
$(5 + 10 - (2 + 6))9$
$(5 + 10 - (8))9$
$(5 + 10 - 8)9$
$(15 - 8)9$
$(7)9$
$7 \cdot 9$
63

(8)
$((6 + 6) \div 2)(2 + 1)$
$((12) \div 2)(2 + 1)$
$(12 \div 2)(2 + 1)$
$(6)(2 + 1)$
$6(2 + 1)$
$6(3)$
$6 \cdot 3$
18

(9)
$(4^2 + (38 - 18)) + 4^2$
$(4^2 + (20)) + 4^2$
$(4^2 + 20) + 4^2$
$(16 + 20) + 4^2$
$(36) + 4^2$
$36 + 4^2$
$36 + 16$
52

(10)
$(50 - (45 - 40)) \div (7 - 2)$
$(50 - (5)) \div (7 - 2)$
$(50 - 5) \div (7 - 2)$
$(45) \div (7 - 2)$
$45 \div (7 - 2)$
$45 \div (5)$
$45 \div 5$
9

Day 42

① $1 - (3^2 \div (14 + 20 - 25))$
$1 - (3^2 \div (34 - 25))$
$1 - (3^2 \div (9))$
$1 - (3^2 \div 9)$
$1 - (9 \div 9)$
$1 - (1)$
$1 - 1$
0

② $5((7^2 + 5 \cdot 3) - 33)$
$5((49 + 5 \cdot 3) - 33)$
$5((49 + 15) - 33)$
$5((64) - 33)$
$5(64 - 33)$
$5(31)$
$5 \cdot 31$
155

③ $(55 - 10) + (1^2 \cdot (23 - 13))$
$(45) + (1^2 \cdot (23 - 13))$
$45 + (1^2 \cdot (23 - 13))$
$45 + (1^2 \cdot (10))$
$45 + (1^2 \cdot 10)$
$45 + (1 \cdot 10)$
$45 + (10)$
$45 + 10$
55

④ $(8 - (1 + 3))^2 \div 8$
$(8 - (4))^2 \div 8$
$(8 - 4)^2 \div 8$
$(4)^2 \div 8$
$4^2 \div 8$
$16 \div 8$
2

⑤ $(6 - 1) + ((1 + 2) + 26)$
$(5) + ((1 + 2) + 26)$
$5 + ((1 + 2) + 26)$
$5 + ((3) + 26)$
$5 + (3 + 26)$
$5 + (29)$
$5 + 29$
34

⑥ $4^2 \div ((12 - 5 \cdot 2) + 6) \cdot 15$
$4^2 \div ((12 - 10) + 6) \cdot 15$
$4^2 \div ((2) + 6) \cdot 15$
$4^2 \div (2 + 6) \cdot 15$
$4^2 \div (8) \cdot 15$
$4^2 \div 8 \cdot 15$
$16 \div 8 \cdot 15$
$2 \cdot 15$
30

⑦ $(80 + (4^2 \div 8)) - 6$
$(80 + (16 \div 8)) - 6$
$(80 + (2)) - 6$
$(80 + 2) - 6$
$(82) - 6$
$82 - 6$
76

⑧ $2(42 + 56 \div (9 - 2))$
$2(42 + 56 \div (7))$
$2(42 + 56 \div 7)$
$2(42 + 8)$
$2(50)$
$2 \cdot 50$
100

⑨ $(6 + (6 - 5))^2 + 8$
$(6 + (1))^2 + 8$
$(6 + 1)^2 + 8$
$(7)^2 + 8$
$7^2 + 8$
$49 + 8$
57

⑩ $2((70 - 55) + 5^2 - 10)$
$2((15) + 5^2 - 10)$
$2(15 + 5^2 - 10)$
$2(15 + 25 - 10)$
$2(40 - 10)$
$2(30)$
$2 \cdot 30$
60

Day 43

① $(5 + (2 \cdot 14) + (5 \cdot 3^2))$
$(5 + (28) + (5 \cdot 3^2))$
$(5 + 28 + (5 \cdot 3^2))$
$(5 + 28 + (5 \cdot 9))$
$(5 + 28 + (45))$
$(5 + 28 + 45)$
$(33 + 45)$
(78)
78

② $6^2 \div ((7 - 3)\,3 - 3)$
$6^2 \div ((4)\,3 - 3)$
$6^2 \div (4 \cdot 3 - 3)$
$6^2 \div (12 - 3)$
$6^2 \div (9)$
$6^2 \div 9$
$36 \div 9$
4

③ $(5 \cdot (48 \div 4))2 - 15$
$(5 \cdot (12))2 - 15$
$(5 \cdot 12)2 - 15$
$(60)2 - 15$
$60 \cdot 2 - 15$
$120 - 15$
105

④ $(56 \div 7)^2 - ((6 \div 6) + 5^2)$
$(8)^2 - ((6 \div 6) + 5^2)$
$8^2 - ((6 \div 6) + 5^2)$
$8^2 - ((1) + 5^2)$
$8^2 - (1 + 5^2)$
$8^2 - (1 + 25)$
$8^2 - (26)$
$8^2 - 26$
$64 - 26$
38

⑤ $100 - 50 + (150 - (50+25))$
$100 - 50 + (150 - (75))$
$100 - 50 + (150 - 75)$
$100 - 50 + (75)$
$100 - 50 + 75$
$50 + 75$
125

⑥ $(22 + (8 \cdot 2^2)) - (4^2 + 16)$
$(22 + (8 \cdot 4)) - (4^2 + 16)$
$(22 + (32)) - (4^2 + 16)$
$(22 + 32) - (4^2 + 16)$
$(54) - (4^2 + 16)$
$54 - (4^2 + 16)$
$54 - (16 + 16)$
$54 - (32)$
$54 - 32$
22

⑦ $(100 - (4^2 + 9 \cdot 8))2$
$(100 - (16 + 9 \cdot 8))2$
$(100 - (16 + 72))2$
$(100 - (88))2$
$(100 - 88)2$
$(12)2$
$12 \cdot 2$
24

⑧ $(3 \cdot 2)^2 - (45 \div (7 - 2))$
$(6)^2 - (45 \div (7 - 2))$
$6^2 - (45 \div (7 - 2))$
$6^2 - (45 \div (5))$
$6^2 - (45 \div 5)$
$6^2 - (9)$
$6^2 - 9$
$36 - 9$
27

⑨ $(4 + (30 - 12)) \div 2$
$(4 + (18)) \div 2$
$(4 + 18) \div 2$
$(22) \div 2$
$22 \div 2$
11

⑩ $2((20 - 14) + 7 - 3^2)$
$2((6) + 7 - 3^2)$
$2(6 + 7 - 3^2)$
$2(6 + 7 - 9)$
$2(13 - 9)$
$2(4)$
$2 \cdot 4$
8

Day 44

① $56 \div ((6 - 2)^2 \div 2)$
$56 \div ((4)^2 \div 2)$
$56 \div (4^2 \div 2)$
$56 \div (16 \div 2)$
$56 \div (8)$
$56 \div 8$
7

② $((3 \cdot 2)^2 - 16) \div 5$
$((6)^2 - 16) \div 5$
$(6^2 - 16) \div 5$
$(36 - 16) \div 5$
$(20) \div 5$
$20 \div 5$
4

③ $23 + ((27 \div 9) \cdot 5^2) - 8^2$
$23 + ((3) \cdot 5^2) - 8^2$
$23 + (3 \cdot 5^2) - 8^2$
$23 + (3 \cdot 25) - 8^2$
$23 + (75) - 8^2$
$23 + 75 - 8^2$
$23 + 75 - 64$
$98 - 64$
34

④ $(150 \div 50) \cdot (4 +(27 - 23))$
$(3) \cdot (4 +(27 - 23))$
$3 \cdot (4 +(27 - 23))$
$3 \cdot (4 +(4))$
$3 \cdot (4 + 4)$
$3 \cdot (8)$
$3 \cdot 8$
24

⑤ $((3^2 - 6) \cdot (4 + 2^2)) - 24$
$((9 - 6) \cdot (4 + 2^2)) - 24$
$((3) \cdot (4 + 2^2)) - 24$
$(3 \cdot (4 + 2^2)) - 24$
$(3 \cdot (4 + 4)) - 24$
$(3 \cdot (8)) - 24$
$(3 \cdot 8) - 24$
$(24) - 24$
$24 - 24$
0

⑥ $6^2 \div (16 \div (3 + 1))$
$6^2 \div (16 \div (4))$
$6^2 \div (16 \div 4)$
$6^2 \div (4)$
$6^2 \div 4$
$36 \div 4$
9

⑦ $(61 + (8 - 6))10$
$(61 + (2))10$
$(61 + 2)10$
$(63)10$
$63 \cdot 10$
630

⑧ $9 \cdot 3 + (80 \div (16 - 14))$
$9 \cdot 3 + (80 \div (2))$
$9 \cdot 3 + (80 \div 2)$
$9 \cdot 3 + (40)$
$9 \cdot 3 + 40$
$27 + 40$
67

⑨ $10^2 + ((32 + 68) \div 4)$
$10^2 + ((100) \div 4)$
$10^2 + (100 \div 4)$
$10^2 + (25)$
$10^2 + 25$
$100 + 25$
125

⑩ $(72 - (4^2 + 4 \cdot 7)) \div 2$
$(72 - (16 + 4 \cdot 7)) \div 2$
$(72 - (16 + 28)) \div 2$
$(72 - (44)) \div 2$
$(72 - 44) \div 2$
$(28) \div 2$
$28 \div 2$
14

Day 45

(1) $44 - 28 + ((8 \div 2)^2 - 13)$
$44 - 28 + ((4)^2 - 13)$
$44 - 28 + (4^2 - 13)$
$44 - 28 + (16 - 13)$
$44 - 28 + (3)$
$44 - 28 + 3$
$16 + 3$
19

(2) $50 - ((12 - 7)^2 + 6)$
$50 - ((5)^2 + 6)$
$50 - (5^2 + 6)$
$50 - (25 + 6)$
$50 - (31)$
$50 - 31$
19

(3) $(10^2 \div (37 - 17)) + 5 \cdot 8$
$(10^2 \div (20)) + 5 \cdot 8$
$(10^2 \div 20) + 5 \cdot 8$
$(100 \div 20) + 5 \cdot 8$
$(5) + 5 \cdot 8$
$5 + 5 \cdot 8$
$5 + 40$
45

(4) $90 + ((3 \div 1 + 6)^2 + 2)$
$90 + ((3 + 6)^2 + 2)$
$90 + ((9)^2 + 2)$
$90 + (9^2 + 2)$
$90 + (81 + 2)$
$90 + (83)$
$90 + 83$
173

(5) $9 \cdot 6 + (8 - (3 + 4))$
$9 \cdot 6 + (8 - (7))$
$9 \cdot 6 + (8 - 7)$
$9 \cdot 6 + (1)$
$9 \cdot 6 + 1$
$54 + 1$
55

(6) $48 \div ((18 - 4^2) + (25 - 15))$
$48 \div ((18 - 16) + (25 - 15))$
$48 \div ((2) + (25 - 15))$
$48 \div (2 + (25 - 15))$
$48 \div (2 + (10))$
$48 \div (2 + 10)$
$48 \div (12)$
$48 \div 12$
4

(7) $((20 + 10 - 2) + 8) \div 6$
$((30 - 2) + 8) \div 6$
$((28) + 8) \div 6$
$(28 + 8) \div 6$
$(36) \div 6$
$36 \div 6$
6

(8) $(6 + (12 + 7))5$
$(6 + (19))5$
$(6 + 19)5$
$(25)5$
$25 \cdot 5$
125

(9) $(7 - 4)\ ((8 + 16) \div 2)$
$(3)\ ((8 + 16) \div 2)$
$3\ ((8 + 16) \div 2)$
$3\ ((24) \div 2)$
$3\ (24 \div 2)$
$3\ (12)$
$3 \cdot 12$
36

(10) $10(2 + (17 - 9))$
$10(2 + (8))$
$10(2 + 8)$
$10(10)$
$10 \cdot 10$
100

Day 46

(1) $2 + (6 + (3 \cdot 2)^2) \div 7$
$2 + (6 + (6)^2) \div 7$
$2 + (6 + 6^2) \div 7$
$2 + (6 + 36) \div 7$
$2 + (42) \div 7$
$2 + 42 \div 7$
$2 + 6$
8

(2) $(12 - (1 + 7))^2 \div 2$
$(12 - (8))^2 \div 2$
$(12 - 8)^2 \div 2$
$(4)^2 \div 2$
$4^2 \div 2$
$16 \div 2$
8

(3) $100 - ((45 + 40) - (4 \cdot 2)^2)$
$100 - ((85) - (4 \cdot 2)^2)$
$100 - (85 - (4 \cdot 2)^2)$
$100 - (85 - (8)^2)$
$100 - (85 - 8^2)$
$100 - (85 - 64)$
$100 - 21$
79

(4) $(14 - 6) - (1 + (27 - 22))$
$(8) - (1 + (27 - 22))$
$8 - (1 + (27 - 22))$
$8 - (1 + (5))$
$8 - (1 + 5)$
$8 - (6)$
$8 - 6$
2

(5) $(7^2 + (24 - 13)) + 3^2$
$(7^2 + (11)) + 3^2$
$(7^2 + 11) + 3^2$
$(49 + 11) + 3^2$
$(60) + 3^2$
$60 + 9$
69

(6) $((15 + 20) \div 5)\ (3 + 4)$
$((35) \div 5)\ (3 + 4)$
$(35 \div 5)\ (3 + 4)$
$(7)\ (3 + 4)$
$7\ (3 + 4)$
$7 \cdot 7$
49

(7) $(3 + 2)^2 - (10 - (9 - 3))^2$
$(5)^2 - (10 - (9 - 3))^2$
$5^2 - (10 - (9 - 3))^2$
$5^2 - (10 - (6))^2$
$5^2 - (10 - 6)^2$
$5^2 - (4)^2$
$5^2 - 4^2$
$25 - 16$
9

(8) $(8 - (3 - 2))^2 + 37$
$(8 - (1))^2 + 37$
$(8 - 1)^2 + 37$
$(7)^2 + 37$
$7^2 + 37$
$49 + 37$
86

(9) $8^2 - 1^2 + (24 \div (7 - 1))$
$8^2 - 1^2 + (24 \div (6))$
$8^2 - 1^2 + (24 \div 6)$
$8^2 - 1^2 + (4)$
$8^2 - 1^2 + 4$
$64 - 1^2 + 4$
$64 - 1 + 4$
$63 + 4$
67

(10) $90 \div ((38 + 41) - 7^2)$
$90 \div ((79) - 7^2)$
$90 \div (79 - 7^2)$
$90 \div (79 - 49)$
$90 \div (30)$
$90 \div 30$
3

Day 47

(1) $(40 - (30 - 8)) \div (7 - 4)$
$(40 - (22)) \div (7 - 4)$
$(40 - 22) \div (7 - 4)$
$(18) \div (7 - 4)$
$18 \div (7 - 4)$
$18 \div (3)$
$18 \div 3$
6

(2) $48 \div (1 \cdot (4^2 - 10))$
$48 \div (1 \cdot (16 - 10))$
$48 \div (1 \cdot (6))$
$48 \div (1 \cdot 6)$
$48 \div (6)$
$48 \div 6$
8

(3) $(5^2 + (3 + 3))^2 - (2 \cdot 2)^2$
$(5^2 + (6))^2 - (2 \cdot 2)^2$
$(5^2 + 6)^2 - (2 \cdot 2)^2$
$(25 + 6)^2 - (2 \cdot 2)^2$
$(31)^2 - (2 \cdot 2)^2$
$31^2 - (2 \cdot 2)^2$
$31^2 - (4)^2$
$31^2 - 4^2$
$961 - 4^2$
$961 - 16$
945

(4) $4 + (43 - (2 \cdot 3)^2 \div 6^2)$
$4 + (43 - (6)^2 \div 6^2)$
$4 + (43 - 6^2 \div 6^2)$
$4 + (43 - 36 \div 6^2)$
$4 + (43 - 36 \div 36)$
$4 + (43 - 1)$
$4 + (42)$
$4 + 42$
46

(5) $((22 - 17) \cdot 3) - (18 \div 9)$
$((5) \cdot 3) - (18 \div 9)$
$(5 \cdot 3) - (18 \div 9)$
$(15) - (18 \div 9)$
$15 - (18 \div 9)$
$15 - (2)$
$15 - 2$
13

(6) $(1 + 2)\ (5 \cdot (5 - 4))$
$(3)\ (5 \cdot (5 - 4))$
$3\ (5 \cdot (5 - 4))$
$3\ (5 \cdot (1))$
$3\ (5 \cdot 1)$
$3\ (5)$
$3 \cdot 5$
15

(7) $4^2 + (5^2 - (20 - 5))$
$4^2 + (5^2 - (15))$
$4^2 + (5^2 - 15)$
$4^2 + (25 - 15)$
$4^2 + (10)$
$4^2 + 10$
$16 + 10$
26

(8) $50 - (10 + 2(10 \div 10))$
$50 - (10 + 2(1))$
$50 - (10 + 2 \cdot 1)$
$50 - (10 + 2)$
$50 - (12)$
$50 - 12$
38

(9) $20 + ((3^2 - 2)^2 - (4 \cdot 2))$
$20 + ((9 - 2)^2 - (4 \cdot 2))$
$20 + ((7)^2 - (4 \cdot 2))$
$20 + (7^2 - (4 \cdot 2))$
$20 + (7^2 - (8))$
$20 + (7^2 - 8)$
$20 + (49 - 8)$
$20 + (41)$
$20 + 41$
61

(10) $(40 - 2(4 + 3^2)) + 34$
$(40 - 2(4 + 9)) + 34$
$(40 - 2(13)) + 34$
$(40 - 2 \cdot 13) + 34$
$(40 - 26) + 34$
$(14) + 34$
$14 + 34$
48

Day 48

(1) $100 - ((3 \cdot 4 \div 2)^2 + 4^2)$
$100 - ((12 \div 2)^2 + 4^2)$
$100 - ((6)^2 + 4^2)$
$100 - (6^2 + 4^2)$
$100 - (36 + 4^2)$
$100 - (36 + 16)$
$100 - (52)$
$100 - 52$
48

(2) $4((5 + 3) \div 2) - 9$
$4((8) \div 2) - 9$
$4(8 \div 2) - 9$
$4(4) - 9$
$4 \cdot 4 - 9$
$16 - 9$
7

(3) $(81 \div (4^2 - (2^2 \cdot 3) + 5))$
$(81 \div (4^2 - (4 \cdot 3) + 5))$
$(81 \div (4^2 - (12) + 5))$
$(81 \div (4^2 - 12 + 5))$
$(81 \div (16 - 12 + 5))$
$(81 \div (4 + 5))$
$(81 \div (9))$
$(81 \div 9)$
(9)
9

(4) $(20 - 12)((3 + 7) \div 2)$
$(8)((3 + 7) \div 2)$
$8((3 + 7) \div 2)$
$8((10) \div 2)$
$8(10 \div 2)$
$8(5)$
$8 \cdot 5$
40

(5) $22 + ((6 \div 2)^2 - (1 + 5))$
$22 + ((3)^2 - (1 + 5))$
$22 + (3^2 - (1 + 5))$
$22 + (3^2 - (6))$
$22 + (3^2 - 6)$
$22 + (9 - 6)$
$22 + (3)$
$22 + 3$
25

(6) $(52 - (13 - 6)) \div (8 + 1^2)$
$(52 - (7)) \div (8 + 1^2)$
$(52 - 7) \div (8 + 1^2)$
$(45) \div (8 + 1^2)$
$45 \div (8 + 1^2)$
$45 \div (8 + 1)$
$45 \div (9)$
$45 \div 9$
5

(7) $3 + (5^2 - (5 + 40 \div 4))$
$3 + (5^2 - (5 + 10))$
$3 + (5^2 - (15))$
$3 + (5^2 - 15)$
$3 + (25 - 15)$
$3 + (10)$
$3 + 10$
13

(8) $(8 - (3 - 2))^2 + 37$
$(8 - (1))^2 + 37$
$(8 - 1)^2 + 37$
$(7)^2 + 37$
$7^2 + 37$
$49 + 37$
86

(9) $24 \div ((6 - 4 + 2) \cdot 2)$
$24 \div ((2 + 2) \cdot 2)$
$24 \div ((4) \cdot 2)$
$24 \div (4 \cdot 2)$
$24 \div (8)$
$24 \div 8$
3

(10) $((1 + 3) \cdot 2)^2 - 34$
$((4) \cdot 2)^2 - 34$
$(4 \cdot 2)^2 - 34$
$(8)^2 - 34$
$8^2 - 34$
$64 - 34$
30

Day 49

(1) $((20 + 70) - (3 + 9 \cdot 8)) \div 3$
$((90) - (3 + 9 \cdot 8)) \div 3$
$(90 - (3 + 9 \cdot 8)) \div 3$
$(90 - (3 + 72)) \div 3$
$(90 - (75)) \div 3$
$(90 - 75) \div 3$
$(15) \div 3$
$15 \div 3$
5

(2) $(5 + 10 - (2 + 6))9$
$(5 + 10 - (8))9$
$(5 + 10 - 8)9$
$(15 - 8)9$
$(7)9$
$7 \cdot 9$
63

(3) $(41 - (14 - 5)) \div (12 - 8)$
$(41 - (9)) \div (12 - 8)$
$(41 - 9) \div (12 - 8)$
$(32) \div (12 - 8)$
$32 \div (12 - 8)$
$32 \div (4)$
$32 \div 4$
8

(4) $(9^2 - (5 + 9 \cdot 7)) + 4^2$
$(9^2 - (5 + 63)) + 4^2$
$(9^2 - (68)) + 4^2$
$(9^2 - 68) + 4^2$
$(81 - 68) + 4^2$
$(13) + 4^2$
$13 + 4^2$
$13 + 16$
29

(5) $(2 \cdot (6 - 1))^2 \div 4$
$(2 \cdot (5))^2 \div 4$
$(2 \cdot 5)^2 \div 4$
$(10)^2 \div 4$
$10^2 \div 4$
$100 \div 4$
25

(6) $(18 + 22) \div (4 \cdot (34 - 24))$
$(40) \div (4 \cdot (34 - 24))$
$40 \div (4 \cdot (34 - 24))$
$40 \div (4 \cdot (10))$
$40 \div (4 \cdot 10)$
$40 \div (40)$
$40 \div 40$
1

(7) $4^2 \div ((5 \cdot 7 + 1) - 20)$
$4^2 \div ((35 + 1) - 20)$
$4^2 \div ((36) - 20)$
$4^2 \div (36 - 20)$
$4^2 \div (16)$
$4^2 \div 16$
$16 \div 16$
1

(8) $((10 - 5) \cdot ((9 \div 3)^2 + 1))$
$((5) \cdot ((9 \div 3)^2 + 1))$
$(5 \cdot ((9 \div 3)^2 + 1))$
$(5 \cdot ((3)^2 + 1))$
$(5 \cdot (3^2 + 1))$
$(5 \cdot (9 + 1))$
$(5 \cdot (10))$
$(5 \cdot 10)$
(50)
50

(9) $2((2 + 2 - 2)^2 \div 2)$
$2((4 - 2)^2 \div 2)$
$2((2)^2 \div 2)$
$2(2^2 \div 2)$
$2(4 \div 2)$
$2(2)$
$2 \cdot 2$
4

(10) $((50 - 5 \cdot 8) + 30)2$
$((50 - 40) + 30)2$
$((10) + 30)2$
$(10 + 30)2$
$(40)2$
$40 \cdot 2$
80

Day 50

(1) $(4 \cdot 5 - (2 + 14))(5 + 2)$
$(4 \cdot 5 - (16))(5 + 2)$
$(4 \cdot 5 - 16)(5 + 2)$
$(20 - 16)(5 + 2)$
$(4)(5 + 2)$
$4(5 + 2)$
$4(7)$
$4 \cdot 7$
28

(2) $7((4^2 + 10 \div 2) - 10)$
$7((16 + 10 \div 2) - 10)$
$7((16 + 5) - 10)$
$7((21) - 10)$
$7(21 - 10)$
$7(11)$
$7 \cdot 11$
77

(3) $120 \div ((6 + 3 \cdot 3)2)$
$120 \div ((6 + 9)2)$
$120 \div ((15)2)$
$120 \div (15 \cdot 2)$
$120 \div (30)$
$120 \div 30$
4

(4) $37 - (6 \div 3 \cdot (3 + 1^2)^2)$
$37 - (6 \div 3 \cdot (3 + 1)^2)$
$37 - (6 \div 3 \cdot (4)^2)$
$37 - (6 \div 3 \cdot 4^2)$
$37 - (6 \div 3 \cdot 16)$
$37 - (2 \cdot 16)$
$37 - (32)$
$37 - 32$
5

(5) $((7 - 5)^2 + (5 \cdot 7)) \div 3$
$((2)^2 + (5 \cdot 7)) \div 3$
$(2^2 + (5 \cdot 7)) \div 3$
$(2^2 + (35)) \div 3$
$(2^2 + 35) \div 3$
$(4 + 35) \div 3$
$(39) \div 3$
$39 \div 3$
13

(6) $(2(42 - 24)) + 8 \div 4$
$(2(18)) + 8 \div 4$
$(2 \cdot 18) + 8 \div 4$
$(36) + 8 \div 4$
$36 + 8 \div 4$
$36 + 2$
38

(7) $81 \cdot 2 - ((6 \cdot 2)^2 + 6)$
$81 \cdot 2 - ((12)^2 + 6)$
$81 \cdot 2 - (12^2 + 6)$
$81 \cdot 2 - (144 + 6)$
$81 \cdot 2 - (150)$
$81 \cdot 2 - 150$
$162 - 150$
12

(8) $43 - ((30 - 9) \div 3 + 2)$
$43 - ((21) \div 3 + 2)$
$43 - (21 \div 3 + 2)$
$43 - (7 + 2)$
$43 - (9)$
$43 - 9$
34

(9) $(9 + 1)^2 - (15 - (7 - 4))$
$(10)^2 - (15 - (7 - 4))$
$10^2 - (15 - (7 - 4))$
$10^2 - (15 - (3))$
$10^2 - (15 - 3)$
$10^2 - (12)$
$10^2 - 12$
$100 - 12$
88

(10) $27 + (12 + 30 \div (3 - 2))$
$27 + (12 + 30 \div (1))$
$27 + (12 + 30 \div 1)$
$27 + (12 + 30)$
$27 + (42)$
$27 + 42$
69

Day 51

① $(100 + 50) \div 2 - 3(5-1)$
$(150) \div 2 - 3(5-1)$
$150 \div 2 - 3(5-1)$
$150 \div 2 - 3(4)$
$150 \div 2 - 3 \cdot 4$
$75 - 3 \cdot 4$
$75 - 12$
63

② $(3 + 5) \cdot 2(40 - 4) \div 3^2$
$(8) \cdot 2(40 - 4) \div 3^2$
$8 \cdot 2(40 - 4) \div 3^2$
$8 \cdot 2(36) \div 3^2$
$8 \cdot 2 \cdot 36 \div 3^2$
$8 \cdot 2 \cdot 36 \div 9$
$16 \cdot 36 \div 9$
$576 \div 9$
64

③ $30 - 6 + 4^2 - 3 \cdot 7 + 10$
$30 - 6 + 16 - 3 \cdot 7 + 10$
$30 - 6 + 16 - 21 + 10$
$24 + 16 - 21 + 10$
$40 - 21 + 10$
$19 + 10$
29

④ $(6^2 - 9 \cdot 3)\ (8 + 2 \div 2)$
$(36 - 9 \cdot 3)\ (8 + 2 \div 2)$
$(36 - 27)\ (8 + 2 \div 2)$
$(9)\ (8 + 2 \div 2)$
$9\ (8 + 2 \div 2)$
$9\ (8 + 1)$
$9\ (9)$
$9 \cdot 9$
81

⑤ $(22 + 8)5 \div 25 - 2 \cdot 2$
$(30)5 \div 25 - 2 \cdot 2$
$30 \cdot 5 \div 25 - 2 \cdot 2$
$150 \div 25 - 2 \cdot 2$
$6 - 2 \cdot 2$
$6 - 4$
2

⑥ $24 \div (40 - 4^2 \cdot 2)\ (3 + 7)$
$24 \div (40 - 16 \cdot 2)\ (3 + 7)$
$24 \div (40 - 32)\ (3 + 7)$
$24 \div (8)\ (3 + 7)$
$24 \div 8\ (3 + 7)$
$24 \div 8\ (10)$
$24 \div 8 \cdot 10$
$3 \cdot 10$
30

⑦ $(37 + 5) \div (3^2 - 4 \cdot 4 \div 8)$
$(42) \div (3^2 - 4 \cdot 4 \div 8)$
$42 \div (3^2 - 4 \cdot 4 \div 8)$
$42 \div (9 - 4 \cdot 4 \div 8)$
$42 \div (9 - 16 \div 8)$
$42 \div (9 - 2)$
$42 \div (7)$
$42 \div 7$
6

⑧ $14 \div 2 + 3 \cdot 8 - 20 \div 4$
$7 + 3 \cdot 8 - 20 \div 4$
$7 + 24 - 20 \div 4$
$7 + 24 - 5$
$7 + 19$
26

Day 52

① $5 - (1 \cdot 2) + 3 \div (2 - 1)$
$5 - (2) + 3 \div (2 - 1)$
$5 - 2 + 3 \div (2 - 1)$
$5 - 2 + 3 \div (1)$
$5 - 2 + 3 \div 1$
$5 - 2 + 3$
$3 + 3$
6

② $4 \div 1 \cdot 3^2 \div (2 + 7) - 3$
$4 \div 1 \cdot 3^2 \div (9) - 3$
$4 \div 1 \cdot 3^2 \div 9 - 3$
$4 \div 1 \cdot 9 \div 9 - 3$
$4 \cdot 9 \div 9 - 3$
$36 \div 9 - 3$
$4 - 3$
1

③ $40 + 27 - (5 \cdot 3) + (8 \cdot 1^2)$
$40 + 27 - (15) + (8 \cdot 1^2)$
$40 + 27 - 15 + (8 \cdot 1^2)$
$40 + 27 - 15 + (8 \cdot 1)$
$40 + 27 - 15 + (8)$
$40 + 27 - 15 + 8$
$67 - 15 + 8$
$52 + 8$
60

④ $(4 \cdot 2 + 2)^2 \div 20 - 1 + 1$
$(8 + 2)^2 \div 20 - 1 + 1$
$(10)^2 \div 20 - 1 + 1$
$10^2 \div 20 - 1 + 1$
$100 \div 20 - 1 + 1$
$5 - 1 + 1$
$4 + 1$
5

⑤ $9 - 9 + 1 \cdot 2\ (4 \cdot 3)^2$
$9 - 9 + 1 \cdot 2\ (12)^2$
$9 - 9 + 1 \cdot 2 \cdot 12^2$
$9 - 9 + 1 \cdot 2 \cdot 144$
$9 - 9 + 1 \cdot 288$
$9 - 9 + 288$
$0 + 288$
288

⑥ $(8 + 7 \cdot 4 - 2) - 3^2 \div 9$
$(8 + 28 - 2) - 3^2 \div 9$
$(36 - 2) - 3^2 \div 9$
$(34) - 3^2 \div 9$
$34 - 3^2 \div 9$
$34 - 9 \div 9$
$34 - 1$
33

⑦ $1 \cdot 6 - 2 + (80 \div 40 + 3)$
$1 \cdot 6 - 2 + (2 + 3)$
$1 \cdot 6 - 2 + (5)$
$1 \cdot 6 - 2 + 5$
$6 - 2 + 5$
$4 + 5$
9

⑧ $7 \div (6 - 5) + 28 \cdot (5 \div 5)^2$
$7 \div (1) + 28 \cdot (5 \div 5)^2$
$7 \div 1 + 28 \cdot (5 \div 5)^2$
$7 \div 1 + 28 \cdot (1)^2$
$7 \div 1 + 28 \cdot 1^2$
$7 \div 1 + 28 \cdot 1$
$7 + 28 \cdot 1$
$7 + 28$
35

Day 53

① $9^2 \cdot 2 \div 81 - 1 + 6^2 - 30$
$81 \cdot 2 \div 81 - 1 + 6^2 - 30$
$81 \cdot 2 \div 81 - 1 + 36 - 30$
$162 \div 81 - 1 + 36 - 30$
$2 - 1 + 36 - 30$
$1 + 36 - 30$
$37 - 30$
7

② $5 + 7 - 8 \cdot 3 \div 24 + 2^2$
$5 + 7 - 8 \cdot 3 \div 24 + 4$
$5 + 7 - 24 \div 24 + 4$
$5 + 7 - 1 + 4$
$12 - 1 + 4$
$11 + 4$
15

③ $13 - (42 - 6 \cdot 7) + 5 \cdot 5$
$13 - (42 - 42) + 5 \cdot 5$
$13 - (0) + 5 \cdot 5$
$13 - 0 + 5 \cdot 5$
$13 - 0 + 25$
$13 + 25$
38

④ $18 \div 9 + 5 \cdot (4 + 4 - 5)^2$
$18 \div 9 + 5 \cdot (8 - 5)^2$
$18 \div 9 + 5 \cdot (3)^2$
$18 \div 9 + 5 \cdot 3^2$
$18 \div 9 + 5 \cdot 9$
$2 + 5 \cdot 9$
$2 + 45$
47

⑤ $37 \div (10 + 11 \cdot 1 + 16 \cdot 1)$
$37 \div (10 + 11 + 16 \cdot 1)$
$37 \div (10 + 11 + 16)$
$37 \div (21 + 16)$
$37 \div (37)$
$37 \div 37$
1

⑥ $(2 \cdot 2)^2 + 5 - 13 \div 13 + 17$
$(4)^2 + 5 - 13 \div 13 + 17$
$4^2 + 5 - 13 \div 13 + 17$
$16 + 5 - 13 \div 13 + 17$
$16 + 5 - 1 + 17$
$21 - 1 + 17$
$20 + 17$
37

⑦ $29 + (7 + 8 - 1 \cdot 1)^2 \div 98$
$29 + (7 + 8 - 1)^2 \div 98$
$29 + (7 + 7)^2 \div 98$
$29 + (14)^2 \div 98$
$29 + 14^2 \div 98$
$29 + 196 \div 98$
$29 + 2$
31

⑧ $(4 - 3)^2 + (9 \div 9)^2 \cdot (1 \cdot 3)$
$(1)^2 + (9 \div 9)^2 \cdot (1 \cdot 3)$
$1^2 + (9 \div 9)^2 \cdot (1 \cdot 3)$
$1^2 + (1)^2 \cdot (1 \cdot 3)$
$1^2 + 1^2 \cdot (1 \cdot 3)$
$1^2 + 1^2 \cdot (3)$
$1^2 + 1^2 \cdot 3$
$1 + 1^2 \cdot 3$
$1 + 1 \cdot 3$
$1 + 3$
4

Day 54

① $(9 + 2)^2 - 6 \div 1 \cdot (35 - 18)$
$(11)^2 - 6 \div 1 \cdot (35 - 18)$
$11^2 - 6 \div 1 \cdot (35 - 18)$
$11^2 - 6 \div 1 \cdot (17)$
$11^2 - 6 \div 1 \cdot 17$
$121 - 6 \div 1 \cdot 17$
$121 - 6 \cdot 17$
$121 - 102$
19

② $8 \cdot 5 + (4 - 2)^2 - 2 \div 1$
$8 \cdot 5 + (2)^2 - 2 \div 1$
$8 \cdot 5 + 2^2 - 2 \div 1$
$8 \cdot 5 + 4 - 2 \div 1$
$40 + 4 - 2 \div 1$
$40 + 4 - 2$
$44 - 2$
42

③ $49 \div 7^2 + 3 - 18 \div (4^2 + 2)$
$49 \div 7^2 + 3 - 18 \div (16 + 2)$
$49 \div 7^2 + 3 - 18 \div (18)$
$49 \div 7^2 + 3 - 18 \div 18$
$49 \div 49 + 3 - 18 \div 18$
$1 + 3 - 18 \div 18$
$1 + 3 - 1$
$4 - 1$
3

④ $28 - 5 \cdot 2 \div 10 + 4 \cdot 5$
$28 - 10 \div 10 + 4 \cdot 5$
$28 - 1 + 4 \cdot 5$
$28 - 1 + 20$
$27 + 20$
47

⑤ $(23 - 24 \div 8 + 10) \cdot 2 - 5$
$(23 - 3 + 10) \cdot 2 - 5$
$(20 + 10) \cdot 2 - 5$
$(30) \cdot 2 - 5$
$30 \cdot 2 - 5$
$60 - 5$
55

⑥ $34 + 15 - (2 \cdot 7 - 1)^2 \div 169$
$34 + 15 - (14 - 1)^2 \div 169$
$34 + 15 - (13)^2 \div 169$
$34 + 15 - 13^2 \div 169$
$34 + 15 - 169 \div 169$
$34 + 15 - 1$
$49 - 1$
48

⑦ $4 \cdot 7 + 4^2 - 195 \div (1^2 + 4)$
$4 \cdot 7 + 4^2 - 195 \div (1 + 4)$
$4 \cdot 7 + 4^2 - 195 \div (5)$
$4 \cdot 7 + 4^2 - 195 \div 5$
$4 \cdot 7 + 16 - 195 \div 5$
$28 + 16 - 195 \div 5$
$28 + 16 - 39$
$44 - 39$
5

⑧ $30 \div 2 - 3 \cdot 2 + 20 \cdot 1$
$15 - 3 \cdot 2 + 20 \cdot 1$
$15 - 6 + 20 \cdot 1$
$15 - 6 + 20$
$9 + 20$
29

Day 55

① $100 \div 10 + (8 \cdot 4) - 36 \div 6$
$100 \div 10 + (32) - 36 \div 6$
$100 \div 10 + 32 - 36 \div 6$
$10 + 32 - 36 \div 6$
$10 + 32 - 6$
$42 - 6$
36

② $(150 + 40 \div 8 - 6 + 8) - 157$
$(150 + 5 - 6 + 8) - 157$
$(155 - 6 + 8) - 157$
$(149 + 8) - 157$
$(157) - 157$
$157 - 157$
0

③ $3 \cdot (2 - 1)^2 + (34 \div 17 - 1^2)$
$3 \cdot (1)^2 + (34 \div 17 - 1^2)$
$3 \cdot 1^2 + (34 \div 17 - 1^2)$
$3 \cdot 1^2 + (34 \div 17 - 1)$
$3 \cdot 1^2 + (2 - 1)$
$3 \cdot 1^2 + (1)$
$3 \cdot 1^2 + 1$
$3 \cdot 1 + 1$
$3 + 1$
4

④ $(50 - 9 \cdot 5) - 64 \div 4^2 + 8^2$
$(50 - 45) - 64 \div 4^2 + 8^2$
$(5) - 64 \div 4^2 + 8^2$
$5 - 64 \div 4^2 + 8^2$
$5 - 64 \div 16 + 8^2$
$5 - 64 \div 16 + 64$
$5 - 4 + 64$
$1 + 64$
65

⑤ $9 - 9 + 1 \cdot 2\,(4 \cdot 3)^2$
$9 - 9 + 1 \cdot 2\,(12)^2$
$9 - 9 + 1 \cdot 2 \cdot 12^2$
$9 - 9 + 1 \cdot 2 \cdot 144$
$9 - 9 + 2 \cdot 144$
$9 - 9 + 288$
$0 + 288$
288

⑥ $7^2 - (26 + 74) \div 20 \cdot (7 \div 1)$
$7^2 - (100) \div 20 \cdot (7 \div 1)$
$7^2 - 100 \div 20 \cdot (7 \div 1)$
$7^2 - 100 \div 20 \cdot (7)$
$7^2 - 100 \div 20 \cdot 7$
$49 - 100 \div 20 \cdot 7$
$49 - 5 \cdot 7$
$49 - 35$
14

⑦ $19 + (8 - 8 \div 8 \cdot 2 - 4)^2$
$19 + (8 - 1 \cdot 2 - 4)^2$
$19 + (8 - 2 - 4)^2$
$19 + (6 - 4)^2$
$19 + (2)^2$
$19 + 2^2$
$19 + 4$
23

⑧ $5 \cdot (36 \div 9 + 9 - 12) \cdot 5$
$5 \cdot (4 + 9 - 12) \cdot 5$
$5 \cdot (13 - 12) \cdot 5$
$5 \cdot (1) \cdot 5$
$5 \cdot 1 \cdot 5$
$5 \cdot 5$
25

Day 56

① $(50 - 30) \cdot (2 + 1)^2 - 81 \div 9$
$(20) \cdot (2 + 1)^2 - 81 \div 9$
$20 \cdot (2 + 1)^2 - 81 \div 9$
$20 \cdot (3)^2 - 81 \div 9$
$20 \cdot 3^2 - 81 \div 9$
$20 \cdot 9 - 81 \div 9$
$180 - 81 \div 9$
$180 - 9$
171

② $3^2 \div 9 + (5 \cdot 3 - 15) \cdot 3$
$3^2 \div 9 + (15 - 15) \cdot 3$
$3^2 \div 9 + (0) \cdot 3$
$3^2 \div 9 + 0 \cdot 3$
$9 \div 9 + 0 \cdot 3$
$1 + 0 \cdot 3$
$1 + 0$
1

③ $26 + 2^2 \cdot 2 - 7^2 \div 49 - 0$
$26 + 4 \cdot 2 - 7^2 \div 49 - 0$
$26 + 4 \cdot 2 - 49 \div 49 - 0$
$26 + 8 - 49 \div 49 - 0$
$26 + 8 - 1 - 0$
$34 - 1 - 0$
$33 - 0$
33

④ $5 \cdot (17 - 25 \div 5) \cdot 1 + 5$
$5 \cdot (17 - 5) \cdot 1 + 5$
$5 \cdot (12) \cdot 1 + 5$
$5 \cdot 12 \cdot 1 + 5$
$60 \cdot 1 + 5$
$60 + 5$
65

⑤ $10 + 10 - (100 \div 10) + (1 \cdot 3)^2$
$10 + 10 - (10) + (1 \cdot 3)^2$
$10 + 10 - 10 + (1 \cdot 3)^2$
$10 + 10 - 10 + (3)^2$
$10 + 10 - 10 + 3^2$
$10 + 10 - 10 + 9$
$20 - 10 + 9$
$10 + 9$
19

⑥ $81 \div 3^2 + 25 \div 5^2 \cdot 18 \div 3$
$81 \div 9 + 25 \div 5^2 \cdot 18 \div 3$
$81 \div 9 + 25 \div 25 \cdot 18 \div 3$
$9 + 25 \div 25 \cdot 18 \div 3$
$9 + 1 \cdot 18 \div 3$
$9 + 18 \div 3$
$9 + 6$
15

⑦ $(6^2 \div 3) \cdot (2^2 + 2) - (4 + 3^2)$
$(36 \div 3) \cdot (2^2 + 2) - (4 + 3^2)$
$(12) \cdot (2^2 + 2) - (4 + 3^2)$
$12 \cdot (2^2 + 2) - (4 + 3^2)$
$12 \cdot (4 + 2) - (4 + 3^2)$
$12 \cdot (6) - (4 + 3^2)$
$12 \cdot 6 - (4 + 3^2)$
$12 \cdot 6 - (4 + 9)$
$12 \cdot 6 - (13)$
$12 \cdot 6 - 13$
$72 - 13$
59

⑧ $(15 - 7) \div 8 \cdot 9 - 9 + 1^2$
$(8) \div 8 \cdot 9 - 9 + 1^2$
$8 \div 8 \cdot 9 - 9 + 1^2$
$8 \div 8 \cdot 9 - 9 + 1$
$1 \cdot 9 - 9 + 1$
$9 - 9 + 1$
$0 + 1$
1

Day 57

① $4 \cdot 4^2 - 64 + 9^2 \div 3^2 - 4$
$4 \cdot 16 - 64 + 9^2 \div 3^2 - 4$
$4 \cdot 16 - 64 + 81 \div 3^2 - 4$
$4 \cdot 16 - 64 + 81 \div 9 - 4$
$64 - 64 + 81 \div 9 - 4$
$64 - 64 + 9 - 4$
$0 + 9 - 4$
$9 - 4$
5

② $22 + 23 \cdot 10^2 \div 100 - 10 + 17$
$22 + 23 \cdot 100 \div 100 - 10 + 17$
$22 + 2300 \div 100 - 10 + 17$
$22 + 23 - 10 + 17$
$45 - 10 + 17$
$35 + 17$
52

③ $(5 - 5)^2 + (3 \cdot 3)^2 \div (3 \cdot 3)$
$(0)^2 + (3 \cdot 3)^2 \div (3 \cdot 3)$
$0^2 + (3 \cdot 3)^2 \div (3 \cdot 3)$
$0^2 + (9)^2 \div (3 \cdot 3)$
$0^2 + 9^2 \div (3 \cdot 3)$
$0^2 + 9^2 \div (9)$
$0^2 + 9^2 \div 9$
$0 + 9^2 \div 9$
$0 + 81 \div 9$
$0 + 9$
9

④ $(36 \div 6 + 8 \cdot 1) \cdot (6 \div 6)$
$(6 + 8 \cdot 1) \cdot (6 \div 6)$
$(6 + 8) \cdot (6 \div 6)$
$(14) \cdot (6 \div 6)$
$14 \cdot (6 \div 6)$
$14 \cdot (1)$
$14 \cdot 1$
14

⑤ $(7 + 5)^2 - (8 \div 2 + 2 \cdot 8)$
$(12)^2 - (8 \div 2 + 2 \cdot 8)$
$12^2 - (8 \div 2 + 2 \cdot 8)$
$12^2 - (4 + 2 \cdot 8)$
$12^2 - (4 + 16)$
$12^2 - (20)$
$12^2 - 20$
$144 - 20$
124

⑥ $9 \div (49 - 40) + (3 + 7 \cdot 2)^2$
$9 \div (9) + (3 + 7 \cdot 2)^2$
$9 \div 9 + (3 + 7 \cdot 2)^2$
$9 \div 9 + (3 + 14)^2$
$9 \div 9 + (17)^2$
$9 \div 9 + 17^2$
$9 \div 9 + 289$
$1 + 289$
290

⑦ $29 - 17 + 3 \cdot 32 \div 96 - 8$
$29 - 17 + 96 \div 96 - 8$
$29 - 17 + 1 - 8$
$12 + 1 - 8$
$13 - 8$
5

⑧ $(4 \cdot 2 - 2)^2 - (6 - 3)^2 + 36$
$(8 - 2)^2 - (6 - 3)^2 + 36$
$(6)^2 - (6 - 3)^2 + 36$
$6^2 - (6 - 3)^2 + 36$
$6^2 - (3)^2 + 36$
$6^2 - 3^2 + 36$
$36 - 3^2 + 36$
$36 - 9 + 36$
$27 + 36$
63

Day 58

① $(5^2 - 5) \cdot 2^2 \div 5 + (7^2 \div 7)$
$(25 - 5) \cdot 2^2 \div 5 + (7^2 \div 7)$
$(20) \cdot 2^2 \div 5 + (7^2 \div 7)$
$20 \cdot 2^2 \div 5 + (7^2 \div 7)$
$20 \cdot 2^2 \div 5 + (49 \div 7)$
$20 \cdot 2^2 \div 5 + (7)$
$20 \cdot 2^2 \div 5 + 7$
$20 \cdot 4 \div 5 + 7$
$80 \div 5 + 7$
$16 + 7$
23

② $(3^2 \cdot 4) + (5^2 - 3^2 + 23) - 0$
$(9 \cdot 4) + (5^2 - 3^2 + 23) - 0$
$(36) + (5^2 - 3^2 + 23) - 0$
$36 + (5^2 - 3^2 + 23) - 0$
$36 + (25 - 3^2 + 23) - 0$
$36 + (25 - 9 + 23) - 0$
$36 + (16 + 23) - 0$
$36 + (39) - 0$
$36 + 39 - 0$
$75 - 0$
75

③ $25 \div 5 + 60 - 5 \cdot 2 + 19$
$5 + 60 - 5 \cdot 2 + 19$
$5 + 60 - 10 + 19$
$65 - 10 + 19$
$55 + 19$
74

④ $36 + (7 - 3)^2 \cdot (10 \div 2) \cdot 1$
$36 + (4)^2 \cdot (10 \div 2) \cdot 1$
$36 + 4^2 \cdot (10 \div 2) \cdot 1$
$36 + 4^2 \cdot (5) \cdot 1$
$36 + 4^2 \cdot 5 \cdot 1$
$36 + 16 \cdot 5 \cdot 1$
$36 + 80 \cdot 1$
$36 + 80$
116

⑤ $48 - 6 \cdot (1 \div 1 + 7 \cdot 1)$
$48 - 6 \cdot (1 + 7 \cdot 1)$
$48 - 6 \cdot (1 + 7)$
$48 - 6 \cdot (8)$
$48 - 6 \cdot 8$
$48 - 48$
0

⑥ $3^2 \cdot 81 \div 3^2 - 26 + 12 - 1^2$
$9 \cdot 81 \div 3^2 - 26 + 12 - 1^2$
$9 \cdot 81 \div 9 - 26 + 12 - 1^2$
$9 \cdot 81 \div 9 - 26 + 12 - 1$
$729 \div 9 - 26 + 12 - 1$
$81 - 26 + 12 - 1$
$55 + 12 - 1$
$67 - 1$
66

⑦ $6 + (18 - 36 \div 6) + 150 - 50$
$6 + (18 - 6) + 150 - 50$
$6 + (12) + 150 - 50$
$6 + 12 + 150 - 50$
$18 + 150 - 50$
$168 - 50$
118

⑧ $30 \div 15 + 7 + 5 \cdot 3^2 \div 3$
$30 \div 15 + 7 + 5 \cdot 9 \div 3$
$2 + 7 + 5 \cdot 9 \div 3$
$2 + 7 + 45 \div 3$
$2 + 7 + 15$
$9 + 15$
24

Day 59

① $2^2 + (26 - 10) \cdot 2^2 \div (49 - 41)$
$2^2 + (16) \cdot 2^2 \div (49 - 41)$
$2^2 + 16 \cdot 2^2 \div (49 - 41)$
$2^2 + 16 \cdot 2^2 \div (8)$
$2^2 + 16 \cdot 2^2 \div 8$
$4 + 16 \cdot 2^2 \div 8$
$4 + 16 \cdot 4 \div 8$
$4 + 64 \div 8$
$4 + 8$
12

② $45 - 5 \cdot 3^2 + 3^2 - 5^2 \div 5$
$45 - 5 \cdot 9 + 3^2 - 5^2 \div 5$
$45 - 5 \cdot 9 + 9 - 5^2 \div 5$
$45 - 5 \cdot 9 + 9 - 25 \div 5$
$45 - 45 + 9 - 25 \div 5$
$45 - 45 + 9 - 5$
$0 + 9 - 5$
$9 - 5$
4

③ $4 \cdot 8 - (18 \div 6) + 5 - 5$
$4 \cdot 8 - (3) + 5 - 5$
$4 \cdot 8 - 3 + 5 - 5$
$32 - 3 + 5 - 5$
$29 + 5 - 5$
$34 - 5$
29

④ $(1 \div 1 \cdot 7)^2 + 13 - 3 + 6$
$(1 \cdot 7)^2 + 13 - 3 + 6$
$(7)^2 + 13 - 3 + 6$
$7^2 + 13 - 3 + 6$
$49 + 13 - 3 + 6$
$62 - 3 + 6$
$59 + 6$
65

⑤ $6^2 + 11 - (6 - 6^2 \div 36) \cdot 3^2$
$6^2 + 11 - (6 - 36 \div 36) \cdot 3^2$
$6^2 + 11 - (6 - 1) \cdot 3^2$
$6^2 + 11 - (5) \cdot 3^2$
$6^2 + 11 - 5 \cdot 3^2$
$36 + 11 - 5 \cdot 3^2$
$36 + 11 - 5 \cdot 9$
$36 + 11 - 45$
$47 - 45$
2

⑥ $15 - (6 \cdot 2 + 3) \div (5 - 2)$
$15 - (12 + 3) \div (5 - 2)$
$15 - (15) \div (5 - 2)$
$15 - 15 \div (5 - 2)$
$15 - 15 \div (3)$
$15 - 15 \div 3$
$15 - 5$
10

⑦ $(3 \div 3) + (4 + 3 \cdot 2)^2 + 2$
$(1) + (4 + 3 \cdot 2)^2 + 2$
$1 + (4 + 3 \cdot 2)^2 + 2$
$1 + (4 + 6)^2 + 2$
$1 + (10)^2 + 2$
$1 + 10^2 + 2$
$1 + 100 + 2$
$101 + 2$
103

⑧ $(5 - 1) \cdot 2 + 2 \div (6 - 4)$
$(4) \cdot 2 + 2 \div (6 - 4)$
$4 \cdot 2 + 2 \div (6 - 4)$
$4 \cdot 2 + 2 \div (2)$
$4 \cdot 2 + 2 \div 2$
$8 + 2 \div 2$
$8 + 1$
9

Day 60

① $(35 - 17) \cdot 1 + (50 + 9 \div 3)$
$(18) \cdot 1 + (50 + 9 \div 3)$
$18 \cdot 1 + (50 + 9 \div 3)$
$18 \cdot 1 + (50 + 3)$
$18 \cdot 1 + (53)$
$18 \cdot 1 + 53$
$18 + 53$
71

② $(6 + 4 - 2)^2 \div (4^2 + 4^2 - 30)$
$(10 - 2)^2 \div (4^2 + 4^2 - 30)$
$(8)^2 \div (4^2 + 4^2 - 30)$
$8^2 \div (4^2 + 4^2 - 30)$
$8^2 \div (16 + 4^2 - 30)$
$8^2 \div (16 + 16 - 30)$
$8^2 \div (32 - 30)$
$8^2 \div (2)$
$8^2 \div 2$
$64 \div 2$
32

③ $49 \div 7 + 11 - 100 \div 10 \cdot 1$
$7 + 11 - 100 \div 10 \cdot 1$
$7 + 11 - 10 \cdot 1$
$7 + 11 - 10$
$18 - 10$
8

④ $8 \cdot 64 \div 4 - 2 \cdot 8 - 24$
$512 \div 4 - 2 \cdot 8 - 24$
$128 - 2 \cdot 8 - 24$
$128 - 16 - 24$
$112 - 24$
88

⑤ $3^2 \cdot (2 - 1 + 2) \cdot (2 \cdot 5)$
$3^2 \cdot (1 + 2) \cdot (2 \cdot 5)$
$3^2 \cdot (3) \cdot (2 \cdot 5)$
$3^2 \cdot 3 \cdot (2 \cdot 5)$
$3^2 \cdot 3 \cdot (10)$
$3^2 \cdot 3 \cdot 10$
$9 \cdot 3 \cdot 10$
$27 \cdot 10$
270

⑥ $(20 + 10) \div 5 \cdot 2 \div (2 - 1)^2$
$(30) \div 5 \cdot 2 \div (2 - 1)^2$
$30 \div 5 \cdot 2 \div (2 - 1)^2$
$30 \div 5 \cdot 2 \div (1)^2$
$30 \div 5 \cdot 2 \div 1^2$
$30 \div 5 \cdot 2 \div 1$
$6 \cdot 2 \div 1$
$12 \div 1$
12

⑦ $32 - 9 \div 3 + 21 \div 7 - 7$
$32 - 3 + 21 \div 7 - 7$
$32 - 3 + 3 - 7$
$29 + 3 - 7$
$32 - 7$
25

⑧ $2 \cdot 5 - 2^2 + 10 - 4 \cdot 2^2$
$2 \cdot 5 - 4 + 10 - 4 \cdot 2^2$
$2 \cdot 5 - 4 + 10 - 4 \cdot 4$
$10 - 4 + 10 - 4 \cdot 4$
$10 - 4 + 10 - 16$
$6 + 10 - 16$
$16 - 16$
0

Day 61

① $10 + (5 - (-8))$
$10 + (5 - -8)$
$10 + (5 + 8)$
$10 + (13)$
$10 + 13$
23

② $((8 \div (-4)) + (-24))$
$((8 \div -4) + (-24))$
$((-2) + (-24))$
$(-2 + (-24))$
$(-2 + -24)$
$(-2 - 24)$
(-26)
-26

③ $((-3) + (-9 + 3))$
$(-3 + (-9 + 3))$
$(-3 + (-6))$
$(-3 + -6)$
$(-3 - 6)$
(-9)
-9

④ $(-7 \cdot (-36 \div -6))$
$(-7 \cdot (6))$
$(-7 \cdot 6)$
(-42)
-42

⑤ $((12 - (-3)) + 6)$
$((12 - -3) + 6)$
$((12 + 3) + 6)$
$((15) + 6)$
$(15 + 6)$
(21)
21

⑥ $((-15 + -7) - 8)$
$((-15 - 7) - 8)$
$((-22) - 8)$
$(-22 - 8)$
(-30)
-30

⑦ $(((-7) + 2) \cdot 9)$
$((-7 + 2) \cdot 9)$
$((-5) \cdot 9)$
$(-5 \cdot 9)$
(-45)
-45

⑧ $(20 \div -5 + (-35))$
$(20 \div -5 + -35)$
$(20 \div -5 - 35)$
$(-4 - 35)$
(-39)
-39

⑨ $(2 \cdot (-24 \div 8))$
$(2 \cdot (-3))$
$(2 \cdot -3)$
(-6)
-6

⑩ $((-6) + (-10) + 8)$
$(-6 + (-10) + 8)$
$(-6 + -10 + 8)$
$(-6 -10 + 8)$
$(-16 + 8)$
(-8)
-8

Day 62

① $3 - ((-5) + 9)$
$3 - (-5 + 9)$
$3 - (4)$
$3 - 4$
-1

② $((-3) - 5) + 6$
$(-3 - 5) + 6$
$(-8) + 6$
$-8 + 6$
-2

③ $((-7) + (-4)) - 2$
$(-7 + (-4)) - 2$
$(-7 + -4) - 2$
$(-7 - 4) - 2$
$(-11) - 2$
$-11 - 2$
-13

④ $(-3) + ((-8) + 12)$
$-3 + ((-8) + 12)$
$-3 + (-8 + 12)$
$-3 + (4)$
$-3 + 4$
1

⑤ $(8 - (-7)) + 5$
$(8 - -7) + 5$
$(8 + 7) + 5$
$(15) + 5$
$15 + 5$
20

⑥ $((-6) + 4) + (-1)$
$(-6 + 4) + (-1)$
$(-2) + (-1)$
$-2 + (-1)$
$-2 + -1$
$-2 - 1$
-3

⑦ $((-3) + 2) + 9$
$(-3 + 2) + 9$
$(-1) + 9$
$-1 + 9$
8

⑧ $(5 - (-4)) + 7$
$(5 - -4) + 7$
$(5 + 4) + 7$
$(9) + 7$
$9 + 7$
16

⑨ $((-10) - 4) + (-8)$
$(-10 - 4) + (-8)$
$(-14) + (-8)$
$-14 + (-8)$
$-14 + -8$
$-14 - 8$
-22

⑩ $3 + ((-5) + 9)$
$3 + (-5 + 9)$
$3 + (4)$
$3 + 4$
7

Day 63

① -20 - (-15 + 10)
-20 - (-5)
-20 - -5
-20 + 5
-15

② (35 - 11) - 23
(24) - 23
24 - 23
1

③ (7) + (-4 + 2)
7 + (-4 + 2)
7 + (-2)
7 + -2
7 - 2
5

④ 18 - (41 - (-31))
18 - (41 - -31)
18 - (41 + 31)
18 - (72)
18 - 72
-54

⑤ ((-16) + (-5)) - 6
(-16 + (-5)) - 6
(-16 + -5) - 6
(-16 - 5) - 6
(-21) - 6
-21 - 6
-27

⑥ ((-25) - 13 - (-12))
(-25 - 13 - (-12))
(-25 - 13 - -12)
(-25 - 13 + 12)
(-38 + 12)
(-26)
-26

⑦ (17 + (-7)) - 3
(17 + -7) - 3
(17 - 7) - 3
(10) – 3
10 - 3
7

⑧ ((-28) + (-18) - 8)
(-28 + (-18) - 8)
(-28 + -18 - 8)
(-28 - 18 - 8)
(-46 - 8)
(-54)
-54

⑨ (3 + (-3) - (-2))
(3 + -3 - (-2))
(3 - 3 - (-2))
(3 - 3 - -2)
(3 - 3 + 2)
(0 + 2)
(2)
2

⑩ (24 - (-15) - 13)
(24 - -15 - 13)
(24 + 15 - 13)
(39 - 13)
(26)
26

Day 64

① (-36 + (-16) - (-6))
(-36 + -16 - (-6))
(-36 - 16 - (-6))
(-36 - 16 - -6)
(-36 - 16 + 6)
(-52 + 6)
(-46)
-46

② (5) + (-8 + (-3))
5 + (-8 + (-3))
5 + (-8 + -3)
5 + (-8 - 3)
5 + (-11)
5 + -11
5 - 11
-6

③ 12 - ((-9) - (-7))
12 - (-9 - (-7))
12 - (-9- -7)
12 - (-9 + 7)
12 - (-2)
12 - -2
12 + 2
14

④ 47 + ((-13) + 4)
47 + (-13 + 4)
47 + (-9)
47 + -9
47 - 9
38

⑤ (1 + (-1) - (-2))
(1 + -1 - (-2))
(1 - 1 - (-2))
(1 - 1 - -2)
(1 - 1 + 2)
(0 + 2)
(2)
2

⑥ (6-7+(-9))
(6-7+-9)
(6-7-9)
(-1-9)
(-10)
-10

⑦ -29 + (14 - (-8))
-29 + (14 - -8)
-29 + (14 + 8)
-29 + (22)
-29 + 22
-7

⑧ (9 - (-3)) - 22
(9 - -3) - 22
(9 + 3) - 22
(12) – 22
12 - 22
-10

⑨ ((-20) + 10 + (-40))
(-20 + 10 + (-40))
(-20 + 10 + -40)
(-20 + 10 - 40)
(-10 - 40)
(-50)
-50

⑩ ((-3) + (-9 + 3))
(-3 + (-9 + 3))
(-3 + (-6))
(-3 + -6)
(-3 - 6)
(-9)
-9

Day 65

① ((8 -11) + (-6))
((-3) + (-6))
(-3 + (-6))
(-3 + -6)
(-3 - 6)
(-9)
-9

② 4 - ((-9) -7)
4 - (-9 -7)
4 - (-16)
4 - -16
4 + 16
20

③ ((-6) + (-10) + 8)
(-6 + (-10) + 8)
(-6 + -10 + 8)
(-6 - 10 + 8)
(-16 + 8)
(-8)
-8

④ (5 · ((-81) ÷ 9))
(5 · (-81 ÷ 9))
(5 · (-9))
(5 · -9)
(-45)
-45

⑤ ((8 ÷ (-4)) + (-24))
((8 ÷ -4) + (-24))
((-2) + (-24))
(-2 + (-24))
(-2 + -24)
(-2 - 24)
(-26)
-26

⑥ (-11 + 21) · (-3)
(10) · (-3)
10 · (-3)
10 · -3
-30

⑦ 5 · ((-49) ÷ (-7))
5 · (-49 ÷ (-7))
5 · (-49 ÷ -7)
5 · (7)
5 · 7
35

⑧ ((-12) ÷ 6) + 39
(-12 ÷ 6) + 39
(-2) + 39
-2 + 39
37

⑨ ((10 ÷ (-2)) · 10)
((10 ÷ -2) · 10)
((-5) · 10)
(-5 · 10)
(-50)
-50

⑩ (12 - 4 · (-4))
(12 - 4 · -4)
(12 - -16)
(12 + 16)
(28)
28

Day 66

① ((9 · (-3)) ÷ (-27))
((9 · -3) ÷ (-27))
((-27) ÷ (-27))
(-27 ÷ (-27))
(-27 ÷ -27)
(1)
1

② (4 ÷ (-2) - 36)
(4 ÷ -2 - 36)
(-2 - 36)
(-38)
-38

③ 18 + ((-2) · (-5))
18 + (-2 · (-5))
18 + (-2 · -5)
18 + (10)
18 + 10
28

④ ((7 - (-9) + 6))
((7 - -9 + 6))
((7 + 9 + 6))
((16 + 6))
((22))
(22)
22

⑤ (-6 · 8 ÷ 12)
(-48 ÷ 12)
(-4)
-4

⑥ 11 - (21 + (-3))
11 - (21 + -3)
11 - (21 - 3)
11 - (18)
11 - 18
-7

⑦ (9 ÷ 3 - (-31))
(9 ÷ 3 - -31)
(9 ÷ 3 + 31)
(3 + 31)
(34)
34

⑧ ((-16) ÷ 8) · 2
(-16 ÷ 8) · 2
(-2) · 2
-2 · 2
-4

⑨ ((-26) + 14 - (-14))
(-26 + 14 - (-14))
(-26 + 14 - -14)
(-26 + 14 + 14)
(-12 + 14)
(2)
2

⑩ (3 + (-6) ÷ (-3))
(3 + -6 ÷ (-3))
(3 - 6 ÷ (-3))
(3 - 6 ÷ -3)
(3 - -2)
(3 + 2)
(5)
5

Day 67

① ((24 - 44) · (-4))
((-20) · (-4))
(-20 · (-4))
(-20 · -4)
(80)
80

② (2 · (-24 ÷ 8))
(2 · (-3))
(2 · -3)
(-6)
-6

③ (20 - 40 + (-25))
(20 - 40 + -25)
(20 - 40 - 25)
(-20 - 25)
(-45)
-45

④ ((-81) ÷ (3-2))
(-81 ÷ (3-2))
(-81 ÷ (1))
(-81 ÷ 1)
(-81)
-81

⑤ ((-7) + (-7) -7)
(-7 + (-7) -7)
(-7 + -7 -7)
(-7 - 7 -7)
(-14 -7)
(-21)
-21

⑥ (8 ÷ (-2)) · 2
(8 ÷ -2) · 2
(-4) · 2
-4 · 2
-8

⑦ (5 · 5 + (-5))
(5 · 5 + -5)
(5 · 5 - 5)
(25 - 5)
(20)
20

⑧ (3 · (-36 ÷ 6))
(3 · (-6))
(3 · -6)
(-18)
-18

⑨ ((-42 ÷ 6) + 15)
((-7) + 15)
(-7 + 15)
(8)
8

⑩ (13 - ((-4) - 37))
(13 - (-4 - 37))
(13 - (-41))
(13 - -41)
(13 + 41)
(54)
54

Day 68

① ((-18 ÷ 9) -7)
((-2) -7)
(-2 -7)
(-9)
-9

② (5 · 2 ÷ -10)
(10 ÷ -10)
(-1)
-1

③ (4 + (-3) · -6)
(4 + -3 · -6)
(4 - -18)
(4 + 18)
(22)
22

④ (20 ÷ -5 + (-35))
(20 ÷ -5 + -35)
(20 ÷ -5 - 35)
(-4 - 35)
(-39)
-39

⑤ ((8 - 2) ÷ -6)
((6) ÷ -6)
(6 ÷ -6)
(-1)
-1

⑥ (-7 · (-36 ÷ -6))
(-7 · (6))
(-7 · 6)
(-42)
-42

⑦ (100 ÷ -50 - 10)
(-2 - 10)
(-12)
-12

⑧ ((17 + (-19)) · 2)
((17 + -19) · 2)
((17 - 19) · 2)
((-2) · 2)
(-2 · 2)
(-4)
-4

⑨ ((3 + 9) - (-39))
((12) - (-39))
(12 - (-39))
(12 - -39)
(12 + 39)
(51)
51

⑩ ((-34) + (-13) + 7)
(-34 + (-13) + 7)
(-34 + -13 + 7)
(-34 - 13 + 7)
(-47 + 7)
(-40)
-40

Day 69

1. (-1 · 5 · 4)
(-5 · 4)
(-20)
-20

2. ((-5) - ((-2) + (-10)))
(-5 - ((-2) + (-10)))
(-5 - (-2 + (-10)))
(-5 - (-2 + -10))
(-5 - (-2 - 10))
(-5 - (-12))
(-5 - -12)
(-5 + 12)
(7)
7

3. ((24 ÷ 6) + (-24))
((4) + (-24))
(4 + (-24))
(4 + -24)
(4 - 24)
(-20)
-20

4. ((-0) - ((-50) ÷ -5))
(0 - ((-50) ÷ -5))
(0 - (-50 ÷ -5))
(0 - (10))
(0 - 10)
(-10)
-10

5. ((-15 + -7) - 8)
((-15 - 7) - 8)
((-22) - 8)
(-22 - 8)
(-30)
-30

6. (37 + 48 + (-17))
(37 + 48 + -17)
(37 + 48 -17)
(85 -17)
(68)
68

7. ((1) ((-14) + (-42)))
(1 ((-14) + (-42)))
(1 (-14 + (-42)))
(1 (-14 + -42))
(1 (-14 - 42))
(1 (-56))
(1 · -56)
-56

8. (10 ÷ ((-6) - 4))
(10 ÷ (-6 - 4))
(10 ÷ (-10))
(10 ÷ -10)
(-1)
-1

9. ((-11) -6) ÷ -17
(-11 - 6) ÷ -17
(-17) ÷ -17
-17 ÷ -17
1

10. (7- (-3) + (-2))
(7- -3 + (-2))
(7 + 3 + (-2))
(7 + 3 + -2)
(7 + 3 - 2)
(10 - 2)
(8)
8

Day 70

1. (((-19) - (-17)) + (-15))
((-19 - (-17)) + (-15))
((-19 - -17) + (-15))
((-19 + 17) + (-15))
((-2) + (-15))
(-2 + (-15))
(-2 + -15)
(-2 - 15)
(-17)
-17

2. ((8 · -4) ÷ -16)
((-32) ÷ -16)
(-32 ÷ -16)
(-2)
2

3. (3 - 5 · (-6))
(3 - 5 · -6)
(3 - -30)
(3 + 30)
(33)
33

4. (((-7) + 2) · 9)
((-7 + 2) · 9)
((-5) · 9)
(-5 · 9)
(-45)
-45

5. ((-1) + ((-6) - 7))
(-1 + ((-6) - 7))
(-1 + (-6 - 7))
(-1 + (-13))
(-1 + -13)
(-1 - 13)
(-14)
-14

6. (-40 ÷ 4 · -3)
(-10 · -3)
(30)
30

7. ((12 - (-3)) + 6)
((12 - -3) + 6)
((12 + 3) + 6)
((15) + 6)
(15 + 6)
(21)
21

8. ((9 · -2) + (-8))
((-18) + (-8))
(-18 + (-8))
(-18 + -8)
(-18 - 8)
(-26)
-26

9. ((-0) + ((-29) ÷ 1))
(0 + ((-29) ÷ 1))
(0 + (-29 ÷ 1))
(0 + (-29))
(0 + -29)
(0 - 29)
(-29)
-29

10. ((21 ÷ (7)) - (-21))
((21 ÷ 7) - (-21))
((3) - (-21))
(3 - (-21))
(3 - -21)
(3 + 21)
(24)
24

Day 71

1. $\frac{(35+45\div 3)}{5(8-6)}$

$\frac{(35+15)}{5(2)}$

$\frac{(50)}{5\cdot 2}$

$\frac{50}{10}$

5

2. $\frac{9-32\div 8-2}{5^2-3\cdot 7}$

$\frac{9-4-2}{25-3\cdot 7}$

$\frac{5-2}{25-21}$

$\frac{3}{4}$

3. $\frac{7^2-6\cdot 8}{6+35\div 7}$

$\frac{49-6\cdot 8}{6+5}$

$\frac{49-48}{11}$

$\frac{1}{11}$

4. $\frac{2(40-4)}{3^2}$

$\frac{2(36)}{9}$

$\frac{2\cdot 36}{9}$

$\frac{72}{9}$

8

5. $\frac{2(17-6)}{60\div 5+5\cdot 3}$

$\frac{2(11)}{12+5\cdot 3}$

$\frac{2\cdot 11}{12+15}$

$\frac{22}{27}$

6. $\frac{100-5^2\cdot 3}{4+12\cdot 2}$

$\frac{100-25\cdot 3}{4+24}$

$\frac{100-75}{28}$

$\frac{25}{28}$

7. $\frac{2^2\cdot 2}{2(6^2-4(5+1))}$

$\frac{4\cdot 2}{2(6^2-4(6))}$

$\frac{8}{2(6^2-4\cdot 6)}$

$\frac{8}{2(36-4\cdot 6)}$

$\frac{8}{2(36-24)}$

$\frac{8}{2(12)}$

$\frac{8}{2\cdot 12}$

$\frac{8}{24}$

$\frac{1}{3}$

8. $\frac{11+3\cdot 9-2}{(53-8)\div 5}$

$\frac{11+27-2}{(45)\div 5}$

$\frac{38-2}{45\div 5}$

$\frac{36}{9}$

4

Day 72

1. $\frac{3^2+13}{50-5^2}$

$\frac{9+13}{50-25}$

$\frac{22}{25}$

2. $\frac{(22\div 11\cdot 7)+4}{44-(6\cdot 7)}$

$\frac{(2\cdot 7)+4}{44-(42)}$

$\frac{(14)+4}{44-42}$

$\frac{14+4}{44-42}$

$\frac{18}{2}$

9

3. $\frac{8^2\div(7+9)}{5\cdot 3-2^2}$

$\frac{8^2\div(16)}{5\cdot 3-4}$

$\frac{8^2\div 16}{15-4}$

$\frac{64\div 16}{11}$

$\frac{4}{11}$

4. $\frac{11-6+6}{28}$

$\frac{5+6}{28}$

$\frac{11}{28}$

5. $\frac{(13-3)+3^2}{4\cdot 7+8-15}$

$\frac{(10)+3^2}{28+8-15}$

$\frac{10+3^2}{36-15}$

$\frac{10+9}{21}$

$\frac{19}{21}$

6. $\frac{(8^2+(45\div 15))}{(39-29)\cdot 7}$

$\frac{(8^2+(3))}{(10)\cdot 7}$

$\frac{(8^2+3)}{10\cdot 7}$

$\frac{(64+3)}{70}$

$\frac{(67)}{70}$

$\frac{67}{70}$

7. $\frac{(2\cdot 5)-(8\div 4)}{49\div 7^2+7}$

$\frac{(10)-(8\div 4)}{49\div 49+7}$

$\frac{10-(8\div 4)}{1+7}$

$\frac{10-(2)}{1+7}$

$\frac{10-(2)}{8}$

$\frac{10-2}{8}$

$\frac{8}{8}$

1

8. $\frac{24-4^2\div 4+30}{15+(9\cdot 4)}$

$\frac{24-16\div 4+30}{15+(36)}$

$\frac{24-4+30}{15+36}$

$\frac{20+30}{51}$

$\frac{50}{51}$

Day 73

① $\frac{(20 \cdot 2) - 35}{9 + 7 \div 7}$ $\frac{(40) - 35}{9 + 1}$ $\frac{40 - 35}{10}$ $\frac{5}{10}$ $\frac{1}{2}$

② $\frac{6 + (8 - 4) \cdot 6}{10 \div 5 + 8}$ $\frac{6 + (4) \cdot 6}{2 + 8}$ $\frac{6 + 4 \cdot 6}{10}$ $\frac{6 + 24}{10}$ $\frac{30}{10}$ 3

③ $\frac{36 - 10 + 8^2}{15 \cdot 3^2 \div 9}$ $\frac{36 - 10 + 64}{15 \cdot 9 \div 9}$ $\frac{26 + 64}{135 \div 9}$ $\frac{90}{15}$ 6

④ $\frac{4^2 \div 2 \cdot 5 - 7}{(28 - 25 + 7)^2}$ $\frac{16 \div 2 \cdot 5 - 7}{(3 + 7)^2}$ $\frac{8 \cdot 5 - 7}{(10)^2}$ $\frac{40 - 7}{10^2}$ $\frac{33}{100}$

⑤ $\frac{9 \cdot (4 - 2)^2}{(30 \div 15 - 2) + 36}$ $\frac{9 \cdot (2)^2}{(2 - 2) + 36}$ $\frac{9 \cdot 2^2}{(0) + 36}$ $\frac{9 \cdot 4}{0 + 36}$ $\frac{36}{36}$ 1

⑥ $\frac{(16 - 6)^2 \cdot 2}{5 \cdot (24 - 14)}$ $\frac{(10)^2 \cdot 2}{5 \cdot (10)}$ $\frac{10^2 \cdot 2}{5 \cdot 10}$ $\frac{100 \cdot 2}{50}$ $\frac{200}{50}$ 4

⑦ $\frac{(7 + (7 \cdot 3) \div 21)}{((91 - 10) \div 3^2)}$ $\frac{(7 + (21) \div 21)}{((81) \div 3^2)}$ $\frac{(7 + 21 \div 21)}{(81 \div 3^2)}$ $\frac{(7 + 1)}{(81 \div 9)}$ $\frac{(8)}{(9)}$ $\frac{8}{9}$

⑧ $\frac{(30 \div 6 + 2)^2 - 5^2}{3 \cdot 9 - 1}$ $\frac{(5 + 2)^2 - 5^2}{27 - 1}$ $\frac{(7)^2 - 5^2}{27 - 1}$ $\frac{7^2 - 5^2}{26}$ $\frac{49 - 5^2}{26}$ $\frac{49 - 25}{26}$ $\frac{24}{26}$ $\frac{12}{13}$

Day 74

① $\frac{((12 \div 6) \cdot 3)}{90 \div (1 \cdot 10)}$ $\frac{((2) \cdot 3)}{90 \div (10)}$ $\frac{(2 \cdot 3)}{90 \div 10}$ $\frac{(6)}{9}$ $\frac{6}{9}$ $\frac{2}{3}$

② $\frac{25 - 13 + (16 \div 8)}{(15 - 9) \cdot 2^2}$ $\frac{25 - 13 + (2)}{(6) \cdot 2^2}$ $\frac{25 - 13 + 2}{6 \cdot 2^2}$ $\frac{12 + 2}{6 \cdot 4}$ $\frac{14}{24}$ $\frac{7}{12}$

③ $\frac{(4 \cdot 1)^2 \div 2}{5^2 - (8 + 9)}$ $\frac{(4)^2 \div 2}{5^2 - (17)}$ $\frac{4^2 \div 2}{5^2 - 17}$ $\frac{16 \div 2}{25 - 17}$ $\frac{8}{8}$ 1

④ $\frac{18 - 7^2 \div 7 + 3}{8 \cdot 9 \div 2}$ $\frac{18 - 49 \div 7 + 3}{72 \div 2}$ $\frac{18 - 7 + 3}{36}$ $\frac{11 + 3}{36}$ $\frac{14}{36}$ $\frac{7}{18}$

⑤ $\frac{(8 \div 2)^2 - 6}{(8 + 7)2 - 25}$ $\frac{(4)^2 - 6}{(15)2 - 25}$ $\frac{4^2 - 6}{15 \cdot 2 - 25}$ $\frac{16 - 6}{30 - 25}$ $\frac{10}{5}$ 2

⑥ $\frac{(3^2 + 16 - 3^2)}{39 - 10 + 4}$ $\frac{(9 + 16 - 3^2)}{29 + 4}$ $\frac{(9 + 16 - 9)}{33}$ $\frac{(25 - 9)}{33}$ $\frac{(16)}{33}$ $\frac{16}{33}$

⑦ $\frac{19 \cdot 3 - 49 \div 7}{9 \cdot 9 \div 81}$ $\frac{57 - 49 \div 7}{81 \div 81}$ $\frac{57 - 7}{1}$ $\frac{50}{1}$ 50

⑧ $\frac{12 - 10 + 1^2 \cdot 4}{(11 - 7)^2 \cdot 2}$ $\frac{12 - 10 + 1 \cdot 4}{(4)^2 \cdot 2}$ $\frac{12 - 10 + 4}{4^2 \cdot 2}$ $\frac{2 + 4}{16 \cdot 2}$ $\frac{6}{32}$ $\frac{3}{16}$

Day 75

1. $\frac{18 + (5 \cdot 2)^2}{59}$, $\frac{18 + (10)^2}{59}$, $\frac{18 + 10^2}{59}$, $\frac{18 + 100}{59}$, $\frac{118}{59}$, 2

2. $\frac{36 \div 12 - 1}{(3 + 2)13}$, $\frac{3 - 1}{(5)13}$, $\frac{2}{5 \cdot 13}$, $\frac{2}{65}$

3. $\frac{(7^2 + 11)4}{(12 - 10)^2}$, $\frac{(49 + 11)4}{(2)^2}$, $\frac{(60)4}{2^2}$, $\frac{60 \cdot 4}{4}$, $\frac{240}{4}$, 60

4. $\frac{(4 + 5)^2}{(9 - 2) \cdot (8 + 5)}$, $\frac{(9)^2}{(7) \cdot (8 + 5)}$, $\frac{9^2}{7 \cdot (8 + 5)}$, $\frac{81}{7 \cdot (13)}$, $\frac{81}{7 \cdot 13}$, $\frac{81}{91}$

5. $\frac{6 \cdot 3 \div 9}{28 \div (4 - 2)^2 + 2}$, $\frac{18 \div 9}{28 \div (2)^2 + 2}$, $\frac{2}{28 \div 4 + 2}$, $\frac{2}{7 + 2}$, $\frac{2}{9}$

6. $\frac{(18 - (8 + 6))}{49 \div 7}$, $\frac{(18 - (14))}{7}$, $\frac{(18 - 14)}{7}$, $\frac{(4)}{7}$, $\frac{4}{7}$

7. $\frac{8 + ((5 \cdot 5) - 8)}{11 - (7 + 3)}$, $\frac{8 + ((25) - 8)}{11 - (10)}$, $\frac{8 + (25 - 8)}{11 - 10}$, $\frac{8 + (17)}{1}$, $\frac{8 + 17}{1}$, $\frac{25}{1}$, 25

8. $\frac{(9^2 - (4 + 2 \cdot 2)^2)}{5((7 - 3)^2 + 2^2)}$, $\frac{(9^2 - (4 + 4)^2)}{5((4)^2 + 2^2)}$, $\frac{(9^2 - (8)^2)}{5(4^2 + 2^2)}$, $\frac{(9^2 - 8^2)}{5(16 + 2^2)}$, $\frac{(81 - 8^2)}{5(16 + 4)}$, $\frac{(81 - 64)}{5(20)}$, $\frac{(17)}{5 \cdot 20}$, $\frac{17}{100}$

Day 76

1. $\frac{(8 - 6 \div 3)^2}{(2 + 2)^2 \div (2 \cdot 2)}$, $\frac{(8 - 2)^2}{(4)^2 \div (2 \cdot 2)}$, $\frac{(6)^2}{4^2 \div (2 \cdot 2)}$, $\frac{6^2}{4^2 \div (2 \cdot 2)}$, $\frac{36}{4^2 \div (4)}$, $\frac{36}{4^2 \div 4}$, $\frac{36}{16 \div 4}$, $\frac{36}{4}$, 9

2. $\frac{((4^2 \cdot 2) + 7)}{(6^2 + 3)}$, $\frac{((16 \cdot 2) + 7)}{(36 + 3)}$, $\frac{((32) + 7)}{(39)}$, $\frac{(32 + 7)}{39}$, $\frac{(39)}{39}$, $\frac{39}{39}$, 1

3. $\frac{4 \cdot 5 - 6}{17 + 6 - 9 \div 3}$, $\frac{20 - 6}{17 + 6 - 3}$, $\frac{14}{23 - 3}$, $\frac{7}{10}$

4. $\frac{12 \div 3 + 26}{(2 \cdot 5)^2 \div 20}$, $\frac{4 + 26}{(10)^2 \div 20}$, $\frac{30}{10^2 \div 20}$, $\frac{30}{100 \div 20}$, $\frac{30}{100 \div 20}$, $\frac{30}{5}$, 6

5. $\frac{150 - (7 + 4)^2 \cdot 1}{6^2}$, $\frac{150 - (11)^2 \cdot 1}{36}$, $\frac{150 - 11^2 \cdot 1}{36}$, $\frac{150 - 121 \cdot 1}{36}$, $\frac{150 - 121}{36}$, $\frac{29}{36}$

6. $\frac{6(4 - 3)}{3 \div 3(3)}$, $\frac{6(1)}{3 \div 3 \cdot 3}$, $\frac{6 \cdot 1}{1 \cdot 3}$, $\frac{6}{3}$, 2

7. $\frac{8^2}{8^2 \cdot 6 \div 96}$, $\frac{64}{64 \cdot 6 \div 96}$, $\frac{64}{384 \div 96}$, $\frac{64}{4}$, 4

8. $\frac{22 \div 11 \cdot 6 - 10}{0 + 11}$, $\frac{2 \cdot 6 - 10}{11}$, $\frac{12 - 10}{11}$, $\frac{2}{11}$

Day 77

① $\frac{6^2 - 7 \cdot 5}{14 \div 7 \cdot 4^2}$

$\frac{36 - 7 \cdot 5}{14 \div 7 \cdot 16}$

$\frac{36 - 35}{2 \cdot 16}$

$\frac{1}{32}$

② $\frac{8(5) \div 10}{5^2}$

$\frac{8 \cdot 5 \div 10}{25}$

$\frac{40 \div 10}{25}$

$\frac{4}{25}$

③ $\frac{((23 + 19) - 34)}{2 \cdot (16 \div 8)^2}$

$\frac{((42) - 34)}{2 \cdot (2)^2}$

$\frac{(42 - 34)}{2 \cdot 2^2}$

$\frac{(8)}{2 \cdot 4}$

$\frac{8}{8}$

1

④ $\frac{(24 \cdot 3) \div 3^2}{96 \div (2^2 \cdot 6)}$

$\frac{(72) \div 3^2}{96 \div (4 \cdot 6)}$

$\frac{72 \div 3^2}{96 \div 24}$

$\frac{72 \div 9}{96 \div 24}$

$\frac{8}{96 \div 24}$

$\frac{8}{4}$

2

⑤ $\frac{(21 \div (2 + 5))}{41 + 80 \div 10}$

$\frac{(21 \div (7))}{41 + 8}$

$\frac{(21 \div 7)}{49}$

$\frac{(3)}{49}$

$\frac{3}{49}$

⑥ $\frac{(5)1^2}{((25 - 13) + 7)}$

$\frac{5 \cdot 1^2}{((12) + 7)}$

$\frac{5 \cdot 1}{(12 + 7)}$

$\frac{5}{(19)}$

$\frac{5}{19}$

⑦ $\frac{8 \cdot 2^2 - 7^2 \div 49 + 1}{5^2 - 17}$

$\frac{8 \cdot 4 - 7^2 \div 49 + 1}{25 - 17}$

$\frac{8 \cdot 4 - 49 \div 49 + 1}{8}$

$\frac{32 - 49 \div 49 + 1}{8}$

$\frac{32 - 1 + 1}{8}$

$\frac{31 + 1}{8}$

$\frac{32}{8}$

4

⑧ $\frac{(28 - 16) \cdot 8 - 15}{(6 \div 6 + 8)^2}$

$\frac{(12) \cdot 8 - 15}{(1 + 8)^2}$

$\frac{12 \cdot 8 - 15}{(9)^2}$

$\frac{96 - 15}{9^2}$

$\frac{81}{81}$

1

Day 78

① $\frac{21 + (9 \div 3)}{(16 - 13)4}$

$\frac{21 + (3)}{(3)4}$

$\frac{21 + 3}{3 \cdot 4}$

$\frac{24}{12}$

2

② $\frac{(6 \div 3 + 2 - 2)^2}{(5 - 4 \cdot 1)^2}$

$\frac{(2 + 2 - 2)^2}{(5 - 4)^2}$

$\frac{(4 - 2)^2}{(1)^2}$

$\frac{(2)^2}{1^2}$

$\frac{2^2}{1}$

$\frac{4}{1}$

4

③ $\frac{38 + 23 - 7 \cdot 8}{(8)(5)}$

$\frac{38 + 23 - 56}{8(5)}$

$\frac{61 - 56}{8 \cdot 5}$

$\frac{5}{40}$

$\frac{1}{8}$

④ $\frac{20^2 \div 100}{42 + 8^2 - 24}$

$\frac{400 \div 100}{42 + 64 - 24}$

$\frac{4}{106 - 24}$

$\frac{4}{82}$

$\frac{2}{41}$

⑤ $\frac{2^2 \cdot (4^2 + (8 \div 2))}{((9 \div 3) + 3^2) - 2^2}$

$\frac{2^2 \cdot (4^2 + (4))}{((3) + 3^2) - 2^2}$

$\frac{2^2 \cdot (4^2 + 4)}{(3 + 3^2) - 2^2}$

$\frac{2^2 \cdot (16 + 4)}{(3 + 9) - 2^2}$

$\frac{2^2 \cdot (20)}{(12) - 2^2}$

$\frac{2^2 \cdot 20}{12 - 2^2}$

$\frac{4 \cdot 20}{12 - 4}$

$\frac{80}{8}$

10

⑥ $\frac{39 + 7^2 - 49}{9^2 \div 3 + 27}$

$\frac{39 + 49 - 49}{81 \div 3 + 27}$

$\frac{88 - 49}{27 + 27}$

$\frac{39}{54}$

$\frac{13}{18}$

⑦ $\frac{8^2 \cdot 5}{32}$

$\frac{64 \cdot 5}{32}$

$\frac{320}{32}$

10

⑧ $\frac{(12 - 6) + (12 - 6)}{15 \cdot 2 + 5}$

$\frac{(6) + (12 - 6)}{30 + 5}$

$\frac{6 + (12 - 6)}{35}$

$\frac{6 + (6)}{35}$

$\frac{6 + 6}{35}$

$\frac{12}{35}$

Day 79

(1)
$\frac{108 \div 6^2 - 1}{3^2 - 5 + 16}$
$\frac{108 \div 36 - 1}{9 - 5 + 16}$
$\frac{3 - 1}{4 + 16}$
$\frac{2}{20}$
$\frac{1}{10}$

(2)
$\frac{3 \cdot 5 + 2}{((2 + 2)^2 - 8) \cdot 3}$
$\frac{15 + 2}{((4)^2 - 8) \cdot 3}$
$\frac{17}{(4^2 - 8) \cdot 3}$
$\frac{17}{(16 - 8) \cdot 3}$
$\frac{17}{(8) \cdot 3}$
$\frac{17}{8 \cdot 3}$
$\frac{17}{24}$

(3)
$\frac{(4^2 + 11) \div 3^2}{8 \cdot 6^2 - 9^2}$
$\frac{(16 + 11) \div 3^2}{8 \cdot 36 - 9^2}$
$\frac{(27) \div 3^2}{8 \cdot 36 - 81}$
$\frac{27 \div 3^2}{288 - 81}$
$\frac{27 \div 9}{207}$
$\frac{3}{207}$
$\frac{1}{69}$

(4)
$\frac{38 - 33 + 4}{72 \div 8}$
$\frac{5 + 4}{9}$
$\frac{9}{9}$
1

(5)
$\frac{((3 \cdot 8) - 15)}{(24 - (10 \div 5) + 11)}$
$\frac{((24) - 15)}{(24 - (2) + 11)}$
$\frac{(24 - 15)}{(24 - 2 + 11)}$
$\frac{(9)}{(22 + 11)}$
$\frac{9}{(33)}$
$\frac{9}{33}$
$\frac{3}{11}$

(6)
$\frac{20 \div 4 + 20}{(15 - 10)^2 \cdot 5}$
$\frac{5 + 20}{(5)^2 \cdot 5}$
$\frac{5 + 20}{5^2 \cdot 5}$
$\frac{25}{25 \cdot 5}$
$\frac{25}{125}$
$\frac{1}{25}$

(7)
$\frac{(23 - 13) \cdot 6 \div 2}{5 \cdot 2}$
$\frac{(10) \cdot 6 \div 2}{10}$
$\frac{10 \cdot 6 \div 2}{10}$
$\frac{60 \div 2}{10}$
$\frac{30}{10}$
3

(8)
$\frac{1(13 - 3 + 5)}{(18 \div 2)3}$
$\frac{1(10 + 5)}{(9)3}$
$\frac{1(15)}{9 \cdot 3}$
$\frac{1 \cdot 15}{27}$
$\frac{15}{27}$

Day 80

(1)
$\frac{9 - 2 \cdot 4}{14 \div 7 \cdot 4^2}$
$\frac{9 - 8}{14 \div 7 \cdot 16}$
$\frac{1}{2 \cdot 16}$
$\frac{1}{32}$

(2)
$\frac{(3 + 3) \div 1 - 2}{5^2 \div 5 + 3}$
$\frac{(6) \div 1 - 2}{25 \div 5 + 3}$
$\frac{6 \div 1 - 2}{5 + 3}$
$\frac{6 - 2}{8}$
$\frac{4}{8}$
$\frac{1}{2}$

(3)
$\frac{10 + (9 - 5)}{21 \cdot 3}$
$\frac{10 + (4)}{63}$
$\frac{10 + 4}{63}$
$\frac{14}{63}$
$\frac{2}{9}$

(4)
$\frac{19 + 13 - 11}{3^2 \cdot (6 + 4)}$
$\frac{32 - 11}{3^2 \cdot (10)}$
$\frac{21}{3^2 \cdot 10}$
$\frac{21}{9 \cdot 10}$
$\frac{21}{90}$
$\frac{7}{30}$

(5)
$\frac{4^2 + 7^2}{74 + 8 \div 2 - 5}$
$\frac{16 + 7^2}{74 + 4 - 5}$
$\frac{16 + 49}{78 - 5}$
$\frac{65}{73}$

(6)
$\frac{(5 + 1) \cdot (11 - 9)}{10 - 3 + 8}$
$\frac{(6) \cdot (11 - 9)}{7 + 8}$
$\frac{6 \cdot (11 - 9)}{15}$
$\frac{6 \cdot (2)}{15}$
$\frac{6 \cdot 2}{15}$
$\frac{12}{15}$

(7)
$\frac{8 \cdot 2^2 - 14}{3^2}$
$\frac{8 \cdot 4 - 14}{9}$
$\frac{32 - 14}{9}$
$\frac{18}{9}$
2

(8)
$\frac{3 + (4 \cdot 4) \div 2}{28 \div 7 + 10}$
$\frac{3 + (16) \div 2}{4 + 10}$
$\frac{3 + 16 \div 2}{4 + 10}$
$\frac{3 + 8}{4 + 10}$
$\frac{11}{4 + 10}$
$\frac{11}{14}$

Made in United States
Orlando, FL
22 August 2022

21375833R00061